ROBERTS

BIRDS OF KWAZULU-NATAL
AND THEIR
ZULU NAMES

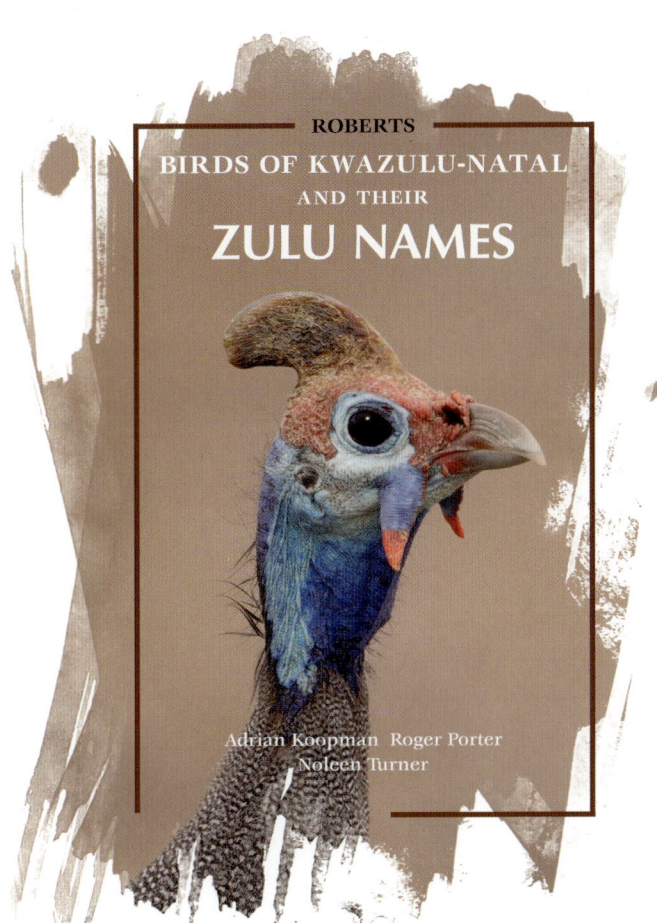

Adrian Koopman Roger Porter
Noleen Turner

This book is DEDICATED to:

The Zulu bird guides who made this publication possible and to all people who are interested in and care for our birds.

"All those who take an interest in the birds they see around them enjoy being able to put a name to them. It is one of the great joys of bird watching and often the first step in becoming interested in their lives and their conservation. Ideally, these names should be in the language most familiar to the individual. Without names to refer to in their mother tongue, it is possible an interest in birds will never take root in the hearts of many."

[Skerrett, Matyot & Rocamora, 2003]

The painting by Angela Beaumont shows the emergence of a Snake Lily (*Scadoxus puniceus*) from the ground at the start of spring. The plant takes its name **idumbe likanhloyile** (the tuber of the Yellow-billed Kite) from this bird, known as **unhloyile**, as both plant and bird appear at the same time of the year. The Zulu month uNcwaba (roughly mid-August to mid-September) has the alternative name uNhloyile, also derived from the name of the kite. These three names – of the bird, the lily, and the month – together show how bird life, plant life and the time of the year are linked in Zulu cultural thought-patterns.

Published by:
The John Voelcker Bird Book Fund

9 Church Square
Cape Town
8001

www.robertsbirds.co.za

Book layout and design Guy Upfold

Artists I B Weiersbye, P Meakin, A Beaumont, G Arnott

Cover Photo Hugh Chittenden

Proofreading Thora Paver

Photographs Hugh Chittenden, Guy Upfold, Ingrid Weiersbye, Graham McGill, Warwick Tarboton, John Carlyon, David Allan, Robert Simmons, Lee Gutteridge, Niel Cillie, Adam Riley, David Weaver, Derek Coley, Lizet Grobbelaar, Johan Grobbelaar and Colleen Downs. All photos not credited are by Hugh Chittenden.

Distribution Jacana Media (www.jacana.co.za)

Printed and bound

Copyright - The authors have copyright over the text contained in the book. The photographers have copyright over their photographs as acknowledged.

1st edition 2020

All rights reserved. No part of this publication may be reproduced, stored in a retrieval system or transmitted, in any form or by any means, without the prior written permission of the copyright holder. However, text extracts of fewer than 100 words will be deemed as fair use and may be reproduced without prior permission as long as such extracts are duly acknowledged.

ISBN 978-1-920602-06-2

CONTENTS

Foreword...2
Authors and the Acknowledgements 3
South Africa's Symbolic National Birds 4
   1. General Introduction...5
      Preamble...5
      Standardisation within the context of name dynamics..................................5
      Birds in Zulu culture and oral literature...6
      Background to terminology development 6
      Standardisation of English and Afrikaans bird names in Southern Africa 7
      Avian nomenclature in Indigenous languages 7
   2. The Zulu Bird Name workshops...8
      Background and involvement of authors 8
      Profiles of four Zulu experts who contributed significantly to the project 10
      Earlier collectors of Zulu bird names 12
      Genus, species and identity...13
      Notion of a 'folk generic' name 13
      Single-word names and expanded names 14
   3. The Zulu Bird Name Workshops 15
      Project methodology 15
      The Workshops 15
      Extending, adapting existing, and creating new names................................16
          (1) Confirmation 16
          (2) Selection and relegation 16
          (3) Redirection..16
          (4) Assignment 16
          (5) Coinage..17
          (6) Transliteration 18
      Sub-division of the bird names in the list..19
      Discussion 20
          (1) Analysis of the species names 20
          (2) Previously recorded in Roberts 20
          (3) Never before recorded in Roberts 20
          (4) Analysis of coined names..20
   4. Conclusion..21
   5. Conservation status of birds in KwaZulu-Natal...22
      Critically endangered species in KwaZulu-Natal 22
      Endangered species in KwaZulu-Natal..22
      Vulnerable species in KwaZulu-Natal..22
      Near-threatened species in KwaZulu-Natal 23
      Extinct bird species in KwaZulu-Natal..23
      South African endemic and near-endemic species occurring in KwaZulu-Natal........23
      Fostering care and an appreciation of our bird heritage..............................23
KwaZulu-Natal Important Birding Areas 24
Crafts inspired by birds 27
Zulu Bird Names..28
Glossary 218
References..221
Notes 222
Index 226

Foreword

In a lecture given some 80 years ago, Austin Roberts expressed his concern that indigenous African bird names were being lost in the process of urbanization. He hoped that his forthcoming book (i.e. the 1st edition of 'Birds of South Africa', 1940) would allow for these bird names to be collected. His reason for including bird names of the various southern African communities and older rural people in his book was to encourage efforts to 'record them more completely and to show how little we really know about them'.

The book *Birds of KwaZulu-Natal and their Zulu Names* by Adrian Koopman, Roger Porter, and Noleen Turner promotes awareness of both the Zulu language and culture as well as, in this case, the province of KwaZulu-Natal's diverse heritage of birds. The book endeavours to bring greater inclusivity to the pleasure of birding and recognises the cultural affiliations that Zulu people have traditionally held with many bird species as reflected in their rich poetic language.

This pioneering publication recognises and embraces the language spoken by more than seven million South Africans, and provides the first standardized reference for bird names in Zulu. The Zulu language is further enriched by providing the names of all 550 bird species that have been recorded in KwaZulu-Natal (except vagrants and pelagic species), of which about 67% of these previously had no Zulu names. Additionally, the book gives insight into the cultural significance of some of the birds found in the province.

The bird names were developed in consultation with bird guides who were trained by BirdLife South Africa. These guides are very knowledgeable birders. Not only do they make a living from guiding both local and international birdwatchers, but they have also become ambassadors for the conservation of the natural environment in the areas where they work and live. They are highly regarded by the people in their communities.

The book will be of value to all people who are interested in birds, including birders, ornithologists and natural scientists employed in the environmental sector. Many of South Africa's birds are unfortunately in a perilous state, with 132 species listed in the regional red data book. This book is a very important educational resource for both building the avitourism industry and for fostering awareness and knowledge that will significantly contribute to the future survival of our country's avifauna.

The authors and the eighteen mother-tongue Zulu-speaking bird experts who assisted with the compilation of *Birds of KwaZulu-Natal and their Zulu Names* are commended for the production of an outstanding and exceptionally useful book.

Mark D. Anderson
Chief Executive Officer
BirdLife South Africa

The Authors

Adrian Koopman is an Emeritus Professor of Zulu at the University of KwaZulu-Natal. His research interests during his academic career have been primarily in onomastics, and he is the author of the *2002 Zulu Names* (which includes a chapter on Zulu bird names), the *2015 Zulu Plant Names*, and the *2019 Zulu Bird Names and Bird Lore*, published by the University of KwaZulu-Natal Press. His over 65 articles and chapters in academic journals and books include the 1990 "Ornimatopoeia: Song reference in English, Afrikaans and Zulu bird names".

Roger Porter has had a life-long interest in birds and has travelled extensively on bird trips in Africa, Madagascar, India and Antarctica. A wildlife ecologist and conservation planner for nearly 39 years in the Natal Parks Board and KZN Wildlife, he was stationed in Hluhluwe-iMfolozi Park and Ithala Game Reserve before retirement. He produced the nomination dossiers for the UNESCO World Heritage inscription of iSimangaliso Wetland Park, and the uKhahlamba Drakensberg Mountains, and has been intimately involved in the World Heritage listing of the Sehlabathebi National Park in Lesotho, the Okavango Delta, extension of Mount Kenya, and the Barberton Mkonjwa World Heritage sites.

Noleen Turner retired after 32 years' service from the African Languages Department at the University of KwaZulu-Natal, Durban. She is currently an Honorary Research Professor at the same institution. Her areas of research include Second Language (Zulu) Teaching, Zulu Oral Studies, Onomastics and Humour Studies. She has published over 40 accredited articles in local and overseas journals. A keen birder, she is the original motivator of the Zulu Bird Names Project described in this book.

Acknowledgements

This Bird Guide would not have been possible were it not for the substantial contributions from the bird guides that participated in six workshops held between 2013 and 2018. These mother-tongue Zulu speakers familiar with the avifauna of KwaZulu-Natal applied their minds, discussed, and unanimously decided on the Zulu name for all 550 species of birds in the province. Particular mention must go to Themba Mthembu, Sakhamuzi Mhlongo, Junior Gabela, Thabile Khuzwayo, and Benson Bhekizenzo Ngubane. In addition thanks go to:

Theo Bukhosini, Siya Dlamini, Abednigo Dube, Jethro Mdlalose, Bheki Mhlongo, Bongani Mthembu, Sakhile Mthenjwa, Daluxolo Ngcobo, Vusi Ngwenya, Phindile Ntshangase, Bheki Nyandeni, Bheki Sithole and Nontuthuko Xaba.

In recognition and with the grateful thanks of the authors, it is fitting that this book is dedicated to them.

We thank Hugh Chittenden, chairperson of the John Voelcker Bird Book Fund (JVBBF) for recognizing the importance of and accepting the manuscript to be published by the JVBBF, also for his advice and guidance on the publication of this book. Ingrid Weiersbye-Porter, member of the Board of Trustees of the JVBBF, liaised between photographers, the authors and board contributing significantly in making valuable suggestions that enriched the publication. Guy Upfold undertook the onerous task of the layout of the book, its design, and the distribution maps of the various species occurring in KwaZulu-Natal.

We are most grateful for the wonderful photographs of the various species of birds found in KwaZulu-Natal that were provided by the following photographers: Hugh Chittenden, Guy Upfold, Ingrid Weiersbye, David Allen, Adam Riley (Rockjumper Birding), Graham McGill, Warwick Tarboton, John Carlyon, Rob Simmons, Lee Gutteridge, Niel Cillie, David Weaver, Derek Coley, Lizet Grobbelaar, Johan Grobbelaar, Garth Batchelor and Colleen Downs.

We thank Heidi Snyman for her drawing of the map of the Important Bird Areas in KwaZulu-Natal.

And finally, we thank our respective spouses: Jewel Koopman, Ingrid Weiersbye-Porter and Sidney Turner for the support, and encouragement over the years of this Zulu Bird Names Project.

SOUTH AFRICA'S SYMBOLIC NATIONAL BIRDS

The Symbolism of the Secretarybird on South Africa's Coat of Arms

The South African Coat Of Arms is the highest visual symbol of the State. It appears on citizens' passports and on their birth, marriage, and death certificates. It also appears at embassies and consulates overseas, and forms part of the Great Seal used to signify the approval of the South African president. The Coat of Arms has many elements organized into two oval groups, one on top of the other. The head and outstretched wings of the Secretarybird (**intinginono**) form the upper part of the top oval. Known for its grace in flight, the Secretarybird on the Coat of Arms acts as a messenger of the heavens, and its open out-stretched wings which the bird uses to defend itself while attacking snakes, simultaneously symbolizes ascendance and protection of the nation from its enemies. Between its wings, the rising sun represents life, knowledge, and the dawning of a new era.

The Secretarybird is a threatened species of conservation concern being listed as Vulnerable to extinction.

The National Bird

The Blue Crane (**indwe**) is the official national bird of South Africa. However its conservation status is given as Near Threatened.

South Africa's national bird, the Blue Crane, which was also depicted on the 5c coins

1. General Introduction

Preamble

This is a book about the Zulu names of birds found in the Zulu-speaking area of South Africa, effectively KwaZulu-Natal and bordering areas. Following the order of the birds in the *Roberts Bird Guide (Chittenden, Davies & Weiersbye, 2016)*, the book lists all the birds found in this region, and gives a standardised Zulu name for each bird. Wherever possible, the underlying meaning or origin of the name is explained. Alternative regional names are given when these are still in common current usage. There are references to Zulu cultural practices relating to certain species of birds, and many of the names are illustrated with related Zulu proverbs and idiomatic expressions.

The names are the result of six years of research workshops conducted with 18 mother-tongue Zulu-speaking bird guides from all corners of KwaZulu-Natal, the workshops being held from 2013 to 2018. The background and the methodology of the workshops is presented, as well as discussing such issues as the need for standardised bird names in Zulu, and the roles played by generic names and species-specific names with vernacular naming.

In recent years the English and Afrikaans names of birds became standardised. This is not to say that some bird names did not change during this period – some, like today's Red-capped Robin-Chat, underwent several name changes. During this period each publication of a bird guide contained names which were accepted at the time of publication as the standardised and 'official' names of the birds of South Africa. Such publications include the *Roberts* series of bird guides, and the different editions of Sinclair and Ryan's *SASOL Birds of Southern Africa* and Newman's *Birds of Southern Africa*. However, the naming of birds in Zulu from the beginning of the 21st century has been incomplete, inaccurate and inconsistent.

This book is the outcome of research that focused on two aspects of Zulu avian nomenclature: firstly, documenting existing oral knowledge and previously-recorded (written) names to preserve this knowledge before it disappears, and to check its accuracy in bird identification; secondly, 'filling in the gaps' where names did not exist for certain clusters of birds, or for individual species within certain clusters. The way in which this was done is described below.

The bird name workshops conducted from 2013 to 2018 with a team of experienced Zulu-speaking bird guides were facilitated by two academic linguists who had previously taught in the African Languages Department at the University of KwaZulu-Natal. The aim of these workshops was to:

- identify and record existing vernacular Zulu names for all birds found in the KwaZulu-Natal region;
- select one commonly used name for a species when two or more names existed;
- distinguish between various species having the same name;
- reassign names where necessary; and
- coin and create new names where none had existed before.

Standardisation within the context of name dynamics

In any given bird guide, both the scientific names and the vernacular bird names, are viewed by readers as the 'standard' or 'official' names of the birds, approved by one or the other 'official' ornithological organizations. Such standardisation is important, as it means that all those concerned with birds – environmental and conservational bodies, professional ornithologists, and birders themselves – all use the same names at any given time. If not, it leads to confusion.

In the main section that follows this introduction, the Zulu name that was regarded by the workshop team as the currently best-known name for any given species of bird is given. In the literature there are other earlier bird names, and also in many cases there are regional differences in these names. However, the name chosen as the standard was the name with the widest geographical currency. In some cases, where two regional names appear to carry equal weight, both were accepted as standard names.

GENERAL INTRODUCTION

Birds in Zulu culture and oral literature

Birds play an important role in the cultural life of the Zulu people, being the focus of many traditional beliefs, as well as featuring strongly in various forms of oral literature, particularly as symbols and metaphors in Zulu praise poetry. Birds may also have their own praise poems, and are furthermore the subject of many a proverb or traditional riddle.

In the traditional belief system birds may play the role of omens and portents, predicting illness, death or warfare, as well as luck, both good and bad. They may also act as heralds, of seasons of the year as well as of times of day. Although traditionally wild birds did not feature strongly on the menu, their feathers were highly sought after for decoration and ornamentation, and were particularly important as military insignia, marking off one regiment from the other.

South African bird guides have normally said little or nothing about the cultural importance of birds, unlike botanical guides, which often give information on how plants may be used for medicinal purposes or as 'magical' charms. Readers interested in the literary or the cultural aspects are directed to Koopman's 2019 *Zulu Bird Names and Bird Lore*, where chapters on the roles played by birds in traditional Zulu oral literature and in traditional Zulu beliefs are given.

Background to terminology development

In 2003, BirdLife South Africa (BLSA) initiated a project to translate *Our Beginners Guide to Birds* (Oliver, 2003) into Zulu. This booklet deals with 187 more commonly encountered birds throughout South Africa. Of the 187 birds listed, over 50 had no Zulu equivalent names documented and many more had shared names. This problem was outlined in a statement made by Maclean (1985, 1993) when he commented that:

"Bird names in the African languages present far more problems than in the European-derived languages. Many of them are generic (i.e. all species of sparrows may have the same name), others are regionally limited in application, one name may be applied to two or more different birds, some well-known birds may have more than one name in a different language, and so on. Most bird species have no African names at all."

This situation led to the current bird naming project initiated in 2012. This was essentially a terminology development project to ensure that for each scientific name in the current BirdLife KwaZulu-Natal bird list, there is to be an equivalent Zulu name.

Terminology and Nomenclature

Most birders generally do not use scientific binomials when referring to birds. They use terms like: 'Yellow-billed Duck', 'Southern Banded Snake-Eagle', *bruinkopvisvanger* (Brown-hooded Kingfisher) and *geelkeelkalkoentjie* (Yellow-throated Longclaw). These names are 'vernacular book names', 'vernacular' implying non-scientific, and 'book' implying names found in publications such as bird guides. These names identify birds at the level of species in the Linnaean taxonomy.

People that are not birders use bird names that identify loose 'clusters' like 'duck' (Zulu **idada**), 'eagle' (Zulu **ukhozi**), and 'dove' (Zulu **ijuba**). These 'clusters' of birds do not coincide exactly with any level in the Linnaean taxonomic hierarchy but are roughly somewhere between 'family' and 'genus'. Since the late eighteenth century ornithologists have divided English cluster names like 'eagle' and 'duck' into separate individually named species following the classification made by taxonomists (Reedman, 2016). It is this species-specific naming that was the goal of the workshops that were held.

This situation of loose-named clusters of birds with the occasional species-specific name such as **intshe** (ostrich) and **indwe** (blue crane) is what caused Maclean to remark that "Bird names in the African languages present far more problems than in the European-derived languages". Maclean's comment is perfectly correct. In a cluster-based system of avian nomenclature such as Zulu there will indeed be many species of birds sharing one name, many species with more than one name, and many species with apparently no name at all.

GENERAL INTRODUCTION

Standardisation of English and Afrikaans vernacular bird names in Southern Africa

Currently in South Africa, the work done by committees and interest groups such as the BirdLife South Africa (BLSA) List Committee and the Afrikaans Bird Name Group, Afrikaanse Voëlnaamgroep (AVNG) has led to there being distinct English and Afrikaans book names for every single species of bird found in Southern Africa.

The Afrikaans Voëlnaamgroep followed a process of extensive consultation, circulating its existence widely in the printed and electronic media and distributing a list of names most commonly used. The Group attempted, wherever possible, to reach decisions through consensus and the main aim has always been not to change names unnecessarily. This process has resulted in the publication of *Roberts Voëlgids (tweedeuitgawe)* in 2018 (Chittenden, H, Davies G & I Weiersbye, 2018). Other principles governing the names on the list include that descriptive names are used whenever possible, and that honorific names such as the English name 'Wahlberg's Eagle' or regional names such as 'Cape Robin-Chat', are only included when no descriptive name can be found. Thus the Afrikaans names for these two birds are, respectively, the *bruin arend* 'brown eagle' and *gewone janfrederik* 'common janfrederik' (an onomatopoeic word for robin-chats). Other Afrikaans bird lists do exist, but the name list provided by the AVNG is used by most authors and publishers of bird books. The Board of BLSA made it policy in 2007 that the AVNG list is accepted as the official Afrikaans list, and only amendments advised by the Group will be accepted.

In terms of the naming of birds in English, this process has not been one which has been exclusively locally compiled. The list is based on the work done by the International Ornithological Committee. The IOC World Bird List is an open access resource of the international community of ornithologists. Their goal is to facilitate worldwide communication in ornithology and conservation based on an up-to-date classification of world birds and a set of English names that follow explicit guidelines for spelling and construction (Gill & Wright 2006). The IOC World Bird List complements three other primary world bird lists that differ slightly in their primary goals and taxonomic philosophy, i.e. *The Clements Checklist of the Birds of the World, The Howard & Moore Complete Checklist of the Birds of the World, 4th Edition*, and *HBW Alive/Bird Life International*. Improved alignment of these independent taxonomic works was discussed at a Round Table discussion by the newly structured International Ornithologists Union, at a 2018 meeting in Vancouver, British Columbia.

Avian Nomenclature in Indigenous Languages

Comprehensive bird lists, however, have not been formulated for bird names for South African Bantu vernacular languages.

Marietta Alberts (2008) outlines the great backlog in the development of African language terminologies and she makes the point that this process which entails extensive field-work, is costly and therefore seldom undertaken. In the specialist language of this branch of linguistics, the Zulu Bird Name Project is described as a 'linguistic community-orientated terminography project', which is to say it is within the discipline of linguistics, has used members of the Zulu-speaking community and is involved in both recording and creating new names for birds.

Ideally, terminology in the particular field of bird names is gleaned from both rural and urban communities for documentation in a central names bank monitored by BLSA. In the case of this project, the information given by the various bird guide experts, much of it orally-held knowledge from their own communities, has been collated and systematised.

2. The Zulu bird name workshops

Background and involvement of authors

Two of the authors, Adrian Koopman and Noleen Turner, have been lecturers in the departments of African Languages, at the erstwhile University of Natal (Koopman) and the erstwhile University of Durban-Westville (Turner). These two universities merged to form the University of KwaZulu-Natal in 2003. After that they continued their academic careers in the same department of African Languages. Both have had similar keen interests in the field of onomastics (science of naming) as well as linguistics and oral literature.

Koopman was working independently on Zulu bird names in the early 1980s. He had been doing field work during this period on Zulu place names, interviewing older Zulu-speaking game guards in different areas under the control of the then Natal Parks Board and the Department of Forestry. Later Koopman used the contacts he had established to interview other older Zulu-speaking members of the Natal Parks Board about bird names. When University of Natal colleague Professor Gordon Maclean, an ornithologist in the Zoology Department, was asked to do the Fifth Edition *of Roberts Birds of South Africa,* Koopman was able to use his dictionary trawling and field research to show Maclean that the Zulu names in previous editions were inconsistent and incorrect for many species. In 1989 Koopman was asked to be a consulting editor for the translation into Zulu of the 1980 *A Beginner's Guide to Our Birds* (Oliver 1980). The 1989 translation, by M. T. Mchunu and B.G. Nkwanyana, was published as *Izinyoni Ezingamashumi Amane Nantathu Zakwela kwaZulu* ('Forty-three Birds from KwaZulu') by uPhiko LweNgcebo Yemvelo KwaZulu, the then governmental wildlife and conservation authority of KwaZulu.

The 1989 Zulu translation seems to have gone unnoticed, because in 2003 BLSA approached Noleen Turner to translate exactly the same book into Zulu. She was also unaware of the existence of the earlier translation, and together with Doris Kumalo it was translated and published in 2003 as *Ibhuku Lokucathulisa Abasaqala Ulwazi Ngezinyoni Zethu* (loosely: 'A book of taking first steps about the knowledge of our birds').

A follow-on project continued in July 2004 with Turner acquiring funding from BLSA to establish contact with expert Zulu-speaking bird guides in several different areas of KwaZulu-Natal in an attempt to try and ascertain whether the missing indigenous names existed but had not been recorded, or whether there were simply no Zulu names known for certain bird species.

Workshop 1: participants, from left to right: Benson Ngubane, Themba Mthembu, Noleen Turner, Sakhile Mathenjwa, Theo Bukhosini, Sakhumuzi Mhlongo.

THE ZULU BIRD NAME WORKSHOPS

Turner's association with both bird names and BirdLife South Africa, through her translation of this beginner's bird guide, led BirdLife South Africa to approach her again in 2011 about the possibilities of a project that would result in an individual Zulu name being recorded for each discrete species of bird occurring in KwaZulu-Natal. As a result of her earlier involvement in the translation of the Oliver book, Turner realised that there were large gaps in the literature where there were either:

- many names for one bird;
- many birds for one name;
- many birds without a name.

She embarked on a project which ultimately required considerable input from mother-tongue Zulu-speakers who were bird experts. Given a grant from the National Research Foundation, Turner set about contacting a number of well-known Zulu-speaking bird guides and game guards from both private and public nature reserves. Her primary contacts were Sakhamuzi Mhlongo, who was based near Richards Bay, Themba Mthembu, who was based at the iSimangaliso Wetlands Park, and Benson Ngubane, a game guard employed at Phinda Forest Lodge. These three experts were present at the 2013 workshop, and they suggested the names of other Zulu-speaking bird experts.

A series of workshops was then proposed to formalise the process which would bring together a group of Zulu-speaking bird guides who indicated their willingness to be involved in the process. Turner contacted Adrian Koopman and invited him to become involved in the project. The first workshop in September 2013 was followed by workshops in 2014, 2015, 2017 and 2018. Funding secured for these was from the NRF Rated Researchers Grant system. The outcome of these workshops was a total revision of the existing Zulu avian nomenclature into a model more closely approximating the English and Afrikaans species-specific names for Southern African birds.

The workshops achieved the following:

- bringing into order the anomalies and contradictions found in the various bird books and dictionaries that have contained Zulu names for birds;
- establishing where one bird has a plethora of names, the most well-known or most commonly used. In such cases, the names adjudged to be less common have been recorded;
- establishing the underlying meanings or connotations of existing bird names;
- clarifying the grammatical constructions behind Zulu bird-name formation, together with the literary devices common to oral literary creations like *izihhasho* (Zulu personal praises);
- establishing names for the genus and for the separate species in a single genus;
- assigning previously recorded but 'unattached' names to specific species;
- coining new names for birds which had no name, either written or in oral usage.

The Zulu bird names recorded in this book are the product of the Zulu-speaking bird experts whose combined knowledge of the Zulu language and the avifauna of KwaZulu-Natal was the determining factor in establishing the names of the different species. A total of 18 Zulu mother-tongue experts attended workshops between 2013 and 2018. They were: Theo Bukhosini, Siya Dlamini, Abednigo Dube, Junior Gabela, Thabile Khuzwayo, Jethro Mdlalose, Bheki Mhlongo, Sakhamuzi Mhlongo, Bongani Mthembu, Themba Mthembu, Sakhile Mthenjwa, Daluxolo Ngcobo, Benson Ngubane, Vusi Ngwenya, Phindile Ntshangase, Bheki Nyandeni, Bheki Sithole and Nontuthuko Xaba.

THE ZULU BIRD NAME WORKSHOPS

Profiles of four Zulu bird experts who contributed significantly to the Zulu Bird Name Project

Sakhamuzi Mhlongo

"I live and breathe birding, it's my passion." Sakhamuzi Mhlongo's words quoted in an article published on May 1 2019 in *SA Country Life* describe a man whose love for birding began as a child in the grasslands of his home village, appropriately called Nyoni (bird) near Amatikulu. After deciding he wanted to work with birds, he trained as a bird guide in 2000 in Wakkerstroom and has been accredited by BirdLifeSA as a guide since 2005.

Mhlongo is a passionate and extremely knowledgeable bird guide who is often contracted by birding tour companies to lead parties of bird watchers intent on finding the 'specials'. Mhlongo's perspectives on birds and bird guiding are given in an editorial on bird guide training in *African Birdlife* (Nov/Dec 2017:4) where he is quoted as saying:

"Birds are important in my life and in my Zulu culture, and I use my knowledge about bird behaviour to educate people in my community."

It is through environmental education that Mhlongo is finding his niche and having an important impact. Over the past decade, he has helped with youth education programmes through the Wildlife and Environment Society of South Africa and he works with the Department of Education to conduct weekly environmental education classes in Eshowe, where he teaches children about forest ecology. He also enjoys mentoring potential future bird guides amongst the youth. He says in an article in *Country Life* (May 2019):

"The people in my village and the nearby reserves now see the value of birds and are supporting bird conservation and creating things like craft to sell."

Mhlongo was awarded one of BirdLifeSA's coveted Owl Awards for his contribution to "giving conservation wings" in 2017.

He has been a stalwart of every one of the workshops held between 2013 and 2018 and has been a lead contributor in the process of revising, adapting and extending the list of Zulu names for birds in KwaZulu-Natal.

Themba Mthembu

Themba Mthembu was born and grew up in northern KwaZulu-Natal in the area known as the Elephant Coast. As a youth he herded cattle on the Nibela Peninsula next to the iSimangaliso Wetland Park World Heritage Site. He has in-depth local knowledge of the Zulu people, their culture, customs and history, as well as the local wildlife and birds. He pursued a career in conservation on leaving school and became an accredited bird guide with BirdLife SA in 2001. Between 2002 and 2004, he worked at Ndumo Game Reserve as an environmental educator for the local schools, and from 2006 to 2007 he upgraded his guiding qualifications to NQF4 (the highest level) also training as a guide in white water canoeing.

In 2008, Mthembu established his own company 'Themba's Birding

THE ZULU BIRD NAME WORKSHOPS

and Ecotours' and in the same year he was accredited as an assessor for the Education, Training, Developing Practices Sector Education and Training Authority, and he was contracted by the Tourism World Academy for tourist-guide training.

Since 2010, Mthembu has been training local bird guides at the Tembe Elephant Park as well as training guides for the Tourism World Academy. This work was recognised in 2015 when he was the recipient of the National Best Tourist Guide of the Year Award – a 'dedicated Premier Tourism Award.' He is the Chairperson of the local uMkhanyakude Tourist Guide Forum and is a registered guide with Tourism KwaZulu-Natal.

Over the past 18 years, Mthembu has dedicated his life to improving the lives of his community and promoting biodiversity conservation. To this end he often helps organise local school meetings to discuss critical environmental issues, including protecting wildlife from poachers and the impact of alien plant species. He is currently working on the 'Birding for Sustainable Living' project, where he has convinced five local homesteads to turn their back gardens into 'bird friendly areas'. In this way he is hoping to promote and improve the livelihoods of people in his local communities and at the same time encouraging them to support tourism and conservation efforts. He has been a recipient of the *Sunday Tribune's* 'Game Changer' Award presented in September of 2018 for his conservation efforts. The Award is given to people who go the extra mile to improve the well-being of their community and environment.

He has recently opened a Centre in May 2019, near Tembe Elephant Park which he has funded himself from the money he has earned through guiding. The Centre focuses mainly on avi-tourism and biodiversity conservation and will be used to host skill development programmes as well as accommodating tourist/school/institution and local community excursions. Mthembu played a highly constructive role in this project with his creativity of thought and ideas at all the workshops that were held.

Junior Gabela

Siphamandla Junior Gabela grew up on the banks of the aMatikulu Estuary in KwaZulu-Natal. His infectious enthusiasm, pleasant and friendly nature, his wealth of knowledge and drive to share his ornithological passion are the qualities that made his business stand out and achieve third place at the Ilembe Chamber's 2017 Entrepreneur Competition. His business 'Birding KwaZulu-Natal with Junior Gabela' offers bird watching, canoeing, environmental education, bird identification courses and team-building. Inspired by Sakhamuzi Mhlongo, his friend with whom he works closely, Junior applied to BirdLife South Africa to undergo training as a bird guide in 2000 and he became accredited in 2005. He is currently working towards his Field Guides Association of Southern Africa level 2 qualifications.

Gabela's interest in understanding the different habitats in coastal KwaZulu-Natal has naturally grown and he has become an environmental educator for the Wildlife and Environmental Society of Southern Africa at their Mtunzini Twinstreams facility. He also assists Marine and Estuarine Research staff with sampling of aquatic organisms and teaches tertiary level ecological courses. In addition, he assists with bird surveys conducted by BirdLife South Africa and he works with students from the Universities of South Africa, Pretoria and KwaZulu-Natal. His main love however, is taking people on bird watching tours, mainly along the KwaZulu-Natal north coast.

In his own words: "Educating and cultivating a love for nature and an interest in environmental research in people, especially the youth, is my goal. If the youth grow up with that passion they will change the future of our environment for the better. I also want to show people the incredible ecotourism potential we have on the North Coast."

THE ZULU BIRD NAME WORKSHOPS

Gabela was awarded one of BirdLife SA's coveted Owl Awards for his contribution to "giving conservation wings" in 2018. He established himself as one of the major contributors to workshop discussions from 2015 and has been one of the three main contributors alongside his mentors Sakhamuzi Mhlongo and Themba Mthembu.

Bhekizenzo Benson Ngubane

Benson Ngubane grew up in Mtubatuba in the Nkundusi area. A school friend inspired him to turn towards the field of biodiversity conservation. During the school holidays he would assist with cleaning up operations in and around the wilderness areas of the iSimangaliso Wetland Park World Heritage Site. After leaving school he joined the army in Mtubatuba then two years later he secured employment at False Bay Park section of Lake St Lucia. Here he checked fences and did general maintenance work in the Park. In 1990 a few members of Benson's family gained employment at the new Phinda Game Reserve. He heard that Phinda was looking for field rangers and was subsequently employed as a tracker in 1991. He completed an advanced tracker's course in 1992 and was employed as a ranger at Mountain Lodge, moving to Forest Lodge in 1993. He also worked at Vlei Lodge where he held the position of Head Ranger for two years.

Ngubane was accredited with Field Guide Association of Southern Africa level 1 in 2002 and has been registered with the Department of Environmental Affairs and Tourism.

Towards the end of 2017 Ngubane took over a new role with the Environmental Education Programme, a collaboration between Mpilonhle & Beyond and the Africa Foundation. The Environmental Education Programme was piloted and launched at Phinda targeting five local schools from the area. Learners in Grades 6 and 7 participate in the programme twice a year during guided tours with Ngubane. As with Mhlongo, Mthembu and Gabela, he has found his niche in the field of education and he focuses not only on birds but on environmental and biodiversity conservation issues as well.

Ngubane contributed to the first bird name workshop in 2013, and to subsequent workshops in 2015 and 2017.

Earlier Collectors of Zulu Bird Names

The first phase of the bird name project was to review the existing literature and to establish exactly what Zulu bird names had previously been recorded for what species. Readers interested in this aspect are referred to Koopman's two articles on the history of Zulu bird names (2018, 2019a) and his book *Zulu Bird Names and Bird Lore* (2019b), where one can find brief outlines of the roles played by French traveller and naturalist Adulphe Delegorgue, Bishop John Colenso, the brothers Richard and John Woodward, the Durban harbour official Harold Bell-Marley, and Father Jacob Gerstner.

Reliability of the data in the dictionaries

A dictionary is by its very nature generally regarded as authoritative, reliable and trustworthy. Yet all dictionaries are compiled by lexicographers, generally academic linguists who may not be fully informed on the many specialist words they include in their dictionary. Clement Doke and Benedict Vilakazi were not ornithologists, and they had to simply accept the bird names submitted to them from various people. It is clear that there was more than just one contributor of Zulu bird name lists and this can be seen, for example, in their entries on certain dove species.

Their entries for **isibhelu**, **ibhobobo** and **isikhombazana** refer to the 'White-breasted Dove', while their entries for **isikhombazanasehlath**i and **inkwabakazana** refer to the 'Tambourine Dove' and the 'Brown Tambourine Dove' respectively. These are all in fact the same bird, as 'White-breasted Dove' is a well-known earlier synonym for the 'Tambourine Dove'. Another example gives two entries, all referring to the same species, today known as the 'Speckled Pigeon' *Columba guinea*: From the explanations of the Zulu names, however, it would appear that *four* birds are involved, each with a distinct English and Latin name:

ijubantendele—Rock Pigeon—*Columba guinea*
ivukuthu—Speckled Rock Pigeon—*Columba arquatrix*
and—Common Rock Pigeon—*Dialiptila phaenota*
and—Olive Pigeon—*Stictoenas arquatrix*

Genus, species and identity

Within each cluster of birds, certain species may have extra salience: a distinctive feature of appearance, habit, or song which marks it as unique within the group. While such a species still belongs to the named cluster 'the folk genus' it may have a specific name as well. Thus the Bateleur is marked as apart from other eagles because of its tumbling acrobatics and its clapping of wings, and while still being an **ukhozi** 'eagle', it is more specifically identified as **ingqungqulu**. The Fish Eagle, too, is unique among eagles for its diet, and while it still belongs to the folk genus **ukhozi**, is specifically named **inkwazi**. In Zulu folk taxonomy, owls are particularly marked by unique and species-specific names. While all owls are **izikhova**, the Pearl-spotted Owlet is known as **inkovana** 'little owl', the White-faced Owl is **umandubulu**, and Verreaux's Eagle-Owl is **ifubesi**. All three species of Eagle-Owl are known as **isikhovampondo** 'owl with horns'. The African Wood Owl not only has its own species-specific name: there are separate names for the male (**umabhengwane**) and the female (**unobathekeli**).

In brief it can be said that the naming of birds in traditional Zulu culture, as with naming in all folk taxonomies, depended entirely on perception: which birds were perceived to be similar and which were perceived to be distinct. There are taxonomic levels in Zulu avian nomenclature just as there are in Linnaean-type taxonomy. Owls are perceived to be part of a wider cluster, perhaps similar in notion to the scientific 'family'; 'owls with horns' are part of a narrower grouping, perhaps akin to the scientific 'genus'; while owlets like the Pearl-spotted and owls like the White-faced have their own unique species names. Zulu taxonomy goes further than scientific naming in giving separately-named status to the male and the female African Wood Owl.

Such a traditional system of avian nomenclature leaves a number of species of birds having no name at all. These species would not be regarded as discrete in a traditional folk taxonomy, but as indistinguishable members of a larger already-named cluster. In many cases where it seems a bird simply should have a traditional Zulu name because it is clearly and obviously distinct from other species in either or both plumage or voice, it was found that these birds did indeed have a traditional Zulu name, but these names had never been recorded in the literature.

Nonetheless, this book provides unique Zulu names for each discrete species of bird which occurs in the Zulu-speaking area and so brings Zulu avian nomenclature into line with English and Afrikaans bird names. To do so of course requires that the notion of what constitutes a species must follow scientific taxonomy and not a traditional folk taxonomy.

Notion of a 'folk generic name'

English words like *eagle*, *duck* and *owl*, and their Afrikaans and Zulu equivalents *arend*, *eend*, *uil*, **ukhozi**, **idada**, and **isikhova** are not the names of species, but the names of clusters. They are linguistic 'building blocks' which can be further qualified to produce a bird name which designates a discrete species. The word *eagle* is not species-specific, but Booted Eagle is. The word **idada** 'duck' is not species-specific, but **idadelimlomophuzi** 'duck with yellow bill' is. Cluster words like **isikhova** and **idada** (and their English and Afrikaans equivalents *owl*, *duck*, *uil* and *eend*) can be called 'folk generics'.

a) some words are already names of clusters of birds:

THE ZULU BIRD NAME WORKSHOPS

Thus the names **ukhozi**, **idada** and **isikhova** are designations of clusters of closely-related birds as opposed to single species. There are many others, for example **inkonjane**, which refers to all swallows and not just one species, and **isigqobhamithi**, a name which covers all woodpeckers.

b) the building block that carries a qualificative:

In the Zulu bird name workshops, some clusters containing several species were found for which there was no name. In such cases a 'folk generic' was coined, with the intention of then further extending it for the different species in the cluster. The following is an example:

There being no previously recorded Zulu name for the three species of mannikin which occur in KwaZulu-Natal, the name **amadojeyana** 'a number of tiny men' was coined, and then this was extended as **amdojeyana ajwayelikile** 'well-known little men' for the Bronze Mannikin, **amadojeyanabomvu** 'little red men' for the Red-backed Mannikin, and **amadojeyanalunga** 'black-and-white little men' for the Magpie Mannikin.

c) generic names:

There are some clusters of birds, with English folk generic names, which do not occur in KwaZulu-Natal. Such clusters were nevertheless given a Zulu folk generic name. For example all pelagic bird clusters were given a generic name even if only few of them occur off the coast of KwaZulu-Natal. There are thus 'generic' names in this book for penguins, albatrosses, shearwaters, petrels, storm petrels and prions. The aim of such a Zulu generic name allows for it to be used in all print media such as newspapers, magazines, text books etc., especially if these are intended to be bilingual (Zulu and English) publications.

In another example, most of the species of kingfisher found in KwaZulu-Natal have one or more previously recorded names. However it was felt that there should be a name that corresponds with the English word kingfisher in its generic sense, and so the name **indwazela** 'one that sits motionlessly staring ahead' was selected.

Single-word names and expanded names

Binomial scientific names have a head noun (the generic name) qualified by a specific epithet. In folk taxonomies most names are single word names. Examples are the 'folk generics' discussed above: the words *eagle*, *duck*, *owl* and *swallow* in English, and their equivalents in Zulu. These are all words found in non-specialist dictionaries. These single word bird names may be extended with some sort of qualifier. These extended names need not necessarily refer to discrete species: *sandgrouse* (as opposed to just 'grouse'), *sparrowhawk*, *moorhen*, *cuckooshrike*, *stormswaël*, *boomvalk*, *baardaasvoël*, *bankduiker* and many more are words which do not necessarily refer to discrete species.

Many of the names coined for species with no previously recorded unique or discrete Zulu name followed a pattern of a 'folk generic' extended with distinguishing qualifications, functioning in the same way as the specific epithet in scientific naming. A good illustration of this is how different species of eagle were named in the workshops, with Verreaux's Eagle being named **ukhozolumnyama** 'black eagle', the Lesser Spotted Eagle being named **ukhozolumabala** 'eagle which has spots' and the African Hawk-Eagle being named **ukhozolumidwayidwa** 'eagle which has many streaks'. For the three species of snake eagle, the first step was to coin the word *indlanyoka* 'what eats snakes', and then to expand this further, creating **indlanyokensundu** 'brown snake-eater' (Brown Snake Eagle), **indlanyokemnyama** 'black snake-eater' (Black-chested Snake Eagle) and **indlanyok empunga** 'grey snake-eater' (Southern Banded Snake Eagle).

Not all new Zulu names coined for birds followed this pattern of building on simple generic terms, and in the main section of this book, there are a considerable number of newly coined bird names which are single words. Examples are **umalusinkomo** 'what herds cattle (Eastern Nicator), **unongoyana** 'little bird from Ngoye' (Green Barbet) and **umavelashone** 'what appears and disappears' (Little Rush Warbler).

3. The Zulu bird name workshops

Overall Project Methodology

Initially a review of the existing literature was undertaken to establish all published Zulu bird names in bird guides, (particularly all Zulu names in *Roberts* editions since the first edition of 1940), and the background to these names. In addition, lists of Zulu names orally accessed and compiled by various people and kept by BirdLife South Africa were also examined. Koopman also compiled a list of Zulu bird names gleaned from scouring various Zulu-English dictionaries, particularly the work of Bryant (1905) and Doke and Vilakazi (1958).

The first three workshops held between 2013 and 2015 dealt with the problem as identified by Maclean of 'many birds – one name; many names – one bird; many birds – no name' and the lack of clarity between folk genus and scientific species.

Then an interrogation of the information gleaned from the first three workshops was done. Roger Porter identified gaps, irregularities, unsuitable names, ensured a comprehensive list of KwaZulu-Natal species excluding rare vagrant birds, informing participants on bird habitats, behaviour and characteristic morphological features, and the lack of reflection of the name in avian reality.

Workshops held in 2017 and 2018 resulted in a final correlation and compilation to form a single master data base of Zulu bird names with linguistic notes, semantic background, and information relevant to particular species of birds. This master data base, in an edited form, constitutes the next part of this book, i.e. the list of Zulu names.

The Workshops

At the first workshop a procedure was established which was followed at all subsequent workshops. Essentially this consisted of a number of steps, which can be summarised as follows:

1) If the bird under consideration had a known name as a species, then confirmation was sought that this name was still in current use. If not, then step two:

2) If a cluster ('folk generic name') existed, but species within the cluster had no individual names, then salient features were sought to make additions to the cluster name such as *omlomophuzi* 'yellow-billed', *emilenzebomvu* 'red-legged' or *waselathi* 'of the forest'. To select an appropriate feature, the field knowledge of the Zulu-speaking experts, together with the information available in the Roberts VII Multimedia program, were used. For each species consideration was taken of general appearance, specific parts of the plumage, individual calls, habitat and diet. If no cluster name existed which could be extended for different species within the cluster, then step three:

3) Where neither cluster name nor species name existed, then a new name had to be coined. To do this, again the salient features of each species had to be determined (flute-like call, reed-dwelling, white rump, and so on). Then the actual grammatical processes could be chosen that would be suitable. The Zulu language has a number of linguistic devices which can be used to create new words, such as the use of the prefixes *–so–*, *–no–* and *–ma–*, the reduplication of noun stems, the compounding of different lexical elements, and the use of literary devices such as metaphor and onomatopoeia.

During these workshop sessions, Turner and Koopman played three roles between them: facilitator, scribe, and operator of the Multimedia. All participants were asked to write on their worksheet what name decisions had been reached for each species after group discussion, and these sheets were returned.

THE ZULU BIRD NAME WORKSHOPS

Extending, adapting existing, and creating new names

The following processes were identified in the analysing of close to 600 Zulu bird names, both old and new: confirmation; selection and relegation; redirection; assignment; coinage; adaptation and extension. These processes are often intertwined and are not always mutually exclusive: for example, adapting and extending existing bird names produces new words in a bird name list, so in a way they are coinages as well.

(1) Confirmation

The most common process, and the one that took the least amount of time and thought, was the process of confirmation. An example is the well-known name **inkwazi** (African Fish Eagle). Every Zulu dictionary records this name for the bird. Research has shown cognates in a number of southern African Bantu languages so the word is old. Confirmation of a name was straightforward if three factors coincided: the name was in the literature, it was well-known by the team, and the name was generally well-known in the wider community.

(2) Selection and relegation

The process of selection operated at both generic and specific levels. At the generic level it was mostly a case of selecting the name of a bird in a particular cluster to function as the name for the whole cluster. An example of this is the case of the roller cluster. Best known among the birds in this cluster is the Lilac-breasted Roller, with the long-established Zulu name **ifefe**. This was selected as the generic name for rollers, and extended appropriately for the European Roller, the Purple Roller and the Broad-billed Roller. The Lilac-breasted became **ifefelihle** 'beautiful roller'.

'Selection', as a process applied to individual species of birds, refers to the choice made when a bird had more than one name, or a single name had been recorded in different forms. The Hadada Ibis, for example, has been equally recorded as **inkankane** and **ihahane**. Clearly these words are both onomatopoeic and describe the bird's raucous call. In terms of selection the choice of the most suitable name was discussed and with more workshop participants being familiar with **inkankane**, this form was selected. The selection of suitable names became more complicated when there were several well-known names for one bird. For example, the Yellow-billed Kite has been recorded as **ukholwe** (with the regional variation **ukholo**), **isikhokhwane** and **unhloyile**. When discussion with the Zulu-speaking bird guides showed that the name **unhloyile** was the most widely-known and widely-used name, it was selected as the standard name for this bird.

Previously-recorded names for birds which are no longer known or no longer in common use are not given in this book. However all earlier known Zulu bird names are given in *Koopman, et al.* (2020).

(3) Redirection

Redirection is in effect a form of name sharing when one species may have multiple names whilst other species in the same folk genus have no names at all. In the case of the Brown-hooded Kingfisher which had the three distinct names **indwazela, unongozolo and unongobotsha,** the first name was redirected to be used as the generic name for all the Kingfishers indicating their behaviour: *ndwaza* in Zulu indicates a motionless position waiting for prey to appear. The second name, **unongozolo**, was redirected to the Striped Kingfisher and then this name was extended to **unongozolwane**. The third name **unongobotsha** was relegated to 'previously known as' and the name **isiphikeleli** was redirected to the Brown-hooded Kingfisher to indicate the repetitive sound of its call.

(4) Assignment

Assignment describes the process whereby unused names listed in dictionary words for identified birds are used for identified species which otherwise would have no Zulu name. For example, the name **unoxhongo**, in Doke and Vilakazi's (1958) dictionary as 'species of heron' was assigned to the Purple Heron (*Ardea purpurea*) which had no previously recorded name. Similarly the Black Crake was assigned the name **isiqhanazana**, an abbreviated version of the name **isiqhananazana**, listed in dictionaries as 'species of water bird'.

THE ZULU BIRD NAME WORKSHOPS

(5) Coinage

The process of coinage can include both adapting and extending existing words in the creating of new names. In addition, original new words can also be coined which have no resemblance to any existing words. Adaptation and extension usually involve the modification of an existing bird name, often used as a generic name. For example, where dictionaries give the meaning of **idada** simply as 'duck', the word was extended to give **idadelibomvu** 'duck which is red' for the South African Shelduck, **idadelimlomophuzi** 'duck which has a yellow bill' for the Yellow-billed Duck, and **idadelincane** 'duck which is small' for the Hottentot Teal.

In the process of creating new bird names there is a manipulation of existing words in the Zulu lexicon which are not bird names. For example, the word *inkawu* 'vervet monkey' is adapted to **isankawu** '[sounds] like a vervet monkey' to create a name for the Southern Pochard.

Onomatopoeic names usually have no resemblance or connection to any previously existing Zulu word. The name **ubhaklakliyo** for the Caspian Tern is a good example: the syllables chosen were meant to emulate the call of the bird.

Adaptation was one of the linguistic strategies used when the Zulu names of Falcons was considered. Maclean (1985) had given **uheshe** as the name of the Lanner and the Peregrine Falcon. After consideration of aspects such as speed, size, habitat and call, it was decided that falcons and buzzards would have the generic name **uklebe,** (previously given to the goshawks), and that **uheshe** would be better suited to hobbies and goshawks. This name was then adapted and extended as **uheshane** as a generic term for the sparrowhawks by adding the diminutive suffix *–ana* indicating their smaller size. These three generic terms were then extended to reveal specific aspects of each species, for example either:

- size as in the use of the diminutive **uheshanyana** for the Little Sparrowhawk; or
- behaviour as in the tendency to congregate in large numbers in the case of **oklebeklebe** the Amur Falcon, the numbers indicated by reduplication of '**klebe**'; or
- markings as in the name **uheshanomidwayidwa** (derived from **imidwayidwa** 'many stripes') for the Shikra.

The processes of adaptation and extension as applied to the smaller raptors:

falcons and buzzards **uklebe** — **oklebeklebe** Amur Falcon
hobbies and goshawks **uhheshe** — **uheshanomidwayidwa** Shikra
sparrowhawks **uhheshane** — **uheshanyana** Little Sparrowhawk

Other interesting types of adaptation arose during the discussion around names for various species of swift. Maclean (1985) had listed three names for the African Black Swift: **ihlabankomo** 'what stabs the head of cattle', **ihlolamvula** 'what predicts the rain' and **ijankomo** a shortened form of **ijiyankomo**, loosely 'what follows the cattle'. Of these three possibilities, the name **ihlolamvula** was selected, as traditional Zulu beliefs do link this bird to the coming of rain. The Little Swift and the Alpine Swift both did not have recorded names. The notion of the coming of rain was adapted to create the names **inhlolazulu** 'what predicts the weather' for the Alpine Swift and **imvuliyeza** 'the rain is coming' for the Little Swift.

Extension was one of the most common methods of forming names for birds which had no previously recorded Zulu names. This was also done by the committees that decided on species-specific English names for South African birds. This process simply takes a generic name like 'eagle' and then qualifies the word 'eagle' with an obvious feature. It was this process of extension which created names like Tawny Eagle, Steppe Eagle, Crowned Eagle and Martial Eagle. The same linguistic strategy of extension using the basic word **ukhozi** was employed in:

- **ukhozimuhlwa** for the Steppe Eagle (*ukhozi* + *umuhlwa* 'termites' – a reference to the bird's diet);
- **ukhozolusisila** for Wahlberg's Eagle (*ukhozi* + *isisila* 'tail' – a reference to the longish tail);
- **ukhozolumabala** for the Lesser Spotted Eagle (*ukhozi* + *amabala* 'spots'); and
- **ukhozolumnyama** for Verreaux's Eagle (ukhozi + –mnyama 'black').

THE ZULU BIRD NAME WORKSHOPS

The coinages that are the most original are those which are not extensions or adaptations of previously existing bird names. These are the most interesting of the new Zulu names for birds. The onomatopoeic names simply try to recreate in combinations of Zulu phonemes the calls and the sounds that the birds make. Many existing names are onomatopoeic in nature: the name **ingududu** for the Southern Ground Hornbill resonates with its *du-du-du-du-du* call; **unohemu** for the Grey Crowned Crane reflects the sound of *ma-hem, ma-hem, mahemu-hemu* in its call. New onomatopoeic names that were coined include **ubhaklakliyo** for the Caspian Tern that refers to the repetition of the rasping sound /kl/ in its raucous call; **umcwicwicwi** for the Green Malkoha, **isipopopo** for the Yellow-rumped Tinkerbird, **iklosi** for the Grey Penduline Tit, **usibó** for the Black-headed Oriole and **isicivó** for the Black-backed Puffback.

Other coinages were adaptations of words that exist in the Zulu lexicon, but not as bird names. Many of these names referred to both the appearance of birds as well as their calls, often using:

- metaphor as in **umakhwaphamnyama** '[the bird with] black armpits' for the Grey Plover and **isipoki** 'spook', which, like the Afrikaans name *spookvoël*, refers to the mournful eerie drawn-out whistle of the Grey-headed Bushshrike; or
- simile as in **isadube** 'like a zebra' for Stierling's Wren-Warbler and **isangulube** 'like a pig' for the Gorgeous Bush Shrike, in reference to its grunting alarm call.

Coinage can also be based on behaviour as in the cleverly named generic term **onozalashiye** 'she that lays and leaves behind' for cuckoos.

(6) Transliteration

Transliteration is a strategy in terminology development where a word in English is adopted into Zulu. When adoption, or 'borrowing' takes place, a word from the source language must conform to the morphological and the phonological systems of the language which adopts it. Thus when Afrikaans *broek* 'trousers' is adopted into Zulu it becomes *ibhulukwe*. The same Afrikaans word *broek* similarly becomes the almost unrecognisable word *marikhoane* in the Sotho name *ntsu-marikhoane* for the Booted Eagle (literally the '*kort-broek* eagle'). And the Swahili name for the same bird is *tai mabuti*, with the English word *boots* becoming *mabuti* when adopted into Swahili and added to *tai* 'eagle'.

As a principle, we avoided 'Zulu-ising' English names for birds. For example, the Zulu name for the Common Moorhen was not given the name 'i-moheni', but rather the word **inkukhuyamanzi** (literally ' the chicken of the water') was coined. In the same way a word like 'idatha' was avoided for the African Darter, rather coining the name **inyoninyoka**, from *inyoni* 'bird' and *inyoka* 'snake', indicating the snake-like shape of its neck. Transliteration does, however, appear in the form of a name given to the African Broadbill: **umasikulufu**. This name is based on the English word *screw* in reference to the bird's circular and horizontal display flight. The transliteration of *screw* here is however part of an original creative process: the bird itself is not known as the 'African Screw' in English.

Coining new names for birds can be seen as a combination of inspiration and linguistic strategy. The inspiration for the creation of these bird names can be found in the distinctive features of the birds such as plumage, diet, behaviour and song. Simultaneously a wide variety of linguistic strategies were adopted. The use of this kind of skill in creating new words and modes of expression, through derivational grammar as well as through literary devices like metaphor and onomatopoeia, is a developed and practised skill normally associated with the professional creators of traditional oral praise poetry – the *izimbongi* (bards) who created and performed praise poems (*izibongo*) in honour of kings and chiefs. One might assume that this oral art form has disappeared in the modern setting of increasingly urbanised society, but the composing of oral poems in honour of the ordinary person, or even of favoured animals, is still evident in Zulu society although it is definitely more prevalent in rural areas. The vast majority of the Zulu bird guides consulted have grown up in the more rural areas of KwaZulu-Natal and indeed still live there, and they continue to practise these age-old skills of word artistry. The stimulating environment in these sessions, in which highly thought-provoking processes were taking place, was both fascinating and mutually motivational. It was a unique experience to be part of the discussion when these extremely creative ideas and thoughts were constantly coming to light.

THE ZULU BIRD NAME WORKSHOPS

Sub-division of the bird names in the List

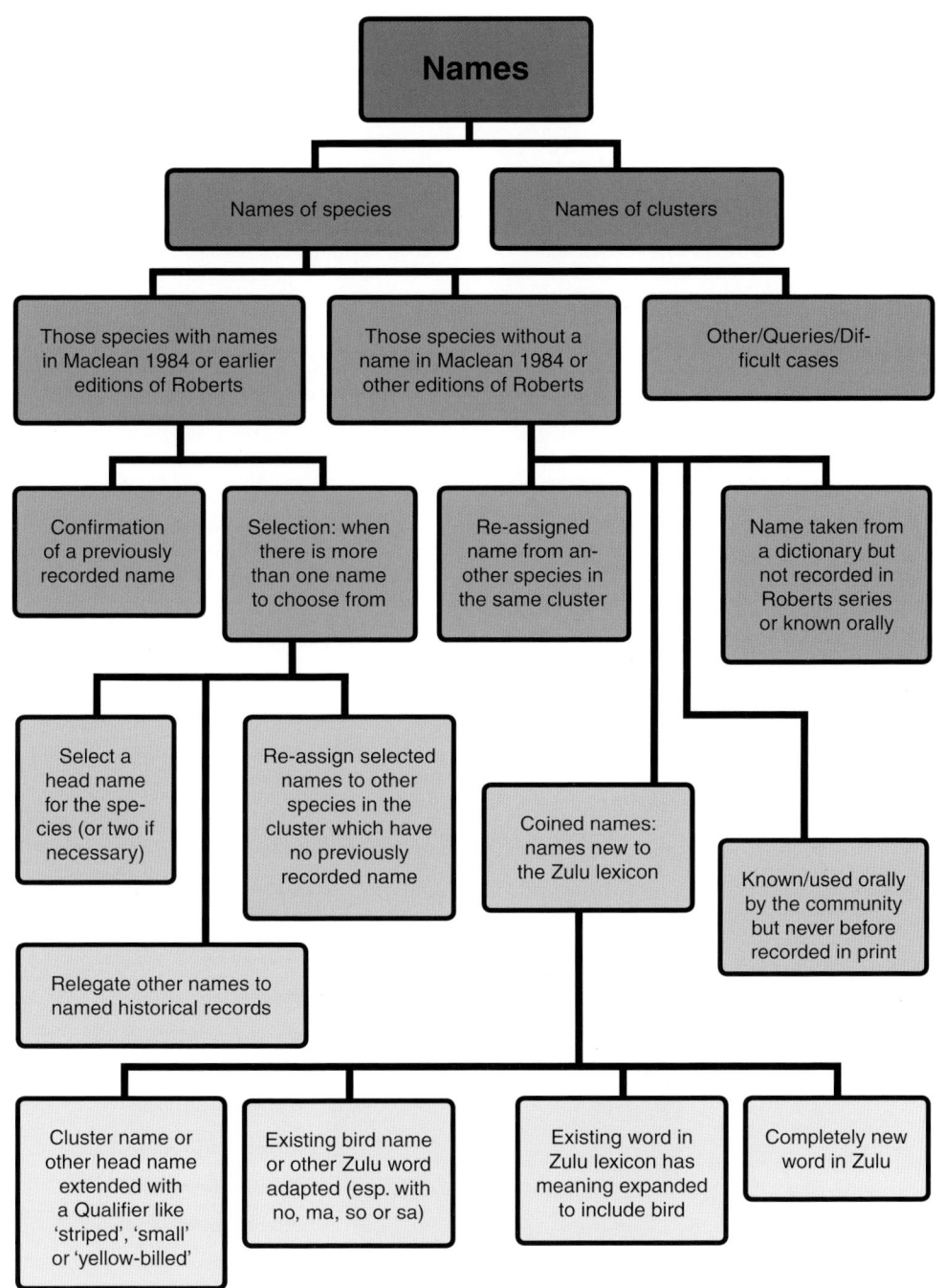

THE ZULU BIRD NAMES PROJECT

Discussion

A total of 550 species of birds are recorded in the list that follows, while a total of 671 names have been analysed. These include the 'historical' names which are not included in this book. It should also be noted that despite the initial aim of 'one bird one name', in some cases two names for the same bird have been given. This has happened when a particular bird is equally well-known by two names, with neither name seen as more widespread than the other.

(1) Analysis of the species names

Of the total of 671 names, 221 names (32,9%) have previously been listed in the Roberts 5th and 6th editions while 446 names (66.5%) are 'new' in the sense that they have never previously been recorded for a bird species. A total of four names were marked 'other–query', accounting for the remaining 0.6% of the names. This means that only a third of the names in the list are previously-known Zulu names for birds. The Zulu Bird Names Project has resulted in a doubling of species-specific names.

(2) Previously recorded in Roberts

Of the 221 Zulu names previously recorded as bird names in the Roberts series, 152 (68.8%) were confirmed as the correct Zulu bird name. A remaining 69 names (31.2%) were either relegated to the historical record as no longer used or well-known, or were re-assigned to other birds, usually within the same cluster.

(3) Never before recorded in Roberts

Of the 446 new bird names in the final list, a majority (373 names, 83.6%) were coined names, i.e. names constructed by adapting other bird names or Zulu nouns which previously did not refer to birds, by compounding different elements to form new bird names, or by other methods of coining. A further 31 names (6.9%) were names confirmed by workshop participants as being widely used orally but never before recorded in print. 29 names (6.5%) were names taken from either Bryant's or Doke and Vilakazi's dictionary but otherwise unrecorded and not in oral use. And then 13 names (2.9%) were names re-assigned from other species which had unused names.

(4) Analysis of the coined names

The largest group of names accounting for well over a half of all the bird names in the List that follows, is the 373 new names that were coined. Most of these (56.3%) consist of those names which use a previously existing name, but extended by qualifying the word in some way.

Other types of coinages are adaptations of existing Zulu words (142 names), some of which are already bird names and some which are not, completely new words in Zulu (4.0%), unrelated to any other known word, and words which previously existed but which have had their meaning expanded to include a bird species, accounting for a mere six names.

Pie-chart showing the relative weighting of the different categories of the total number of Zulu bird names finally decided upon.

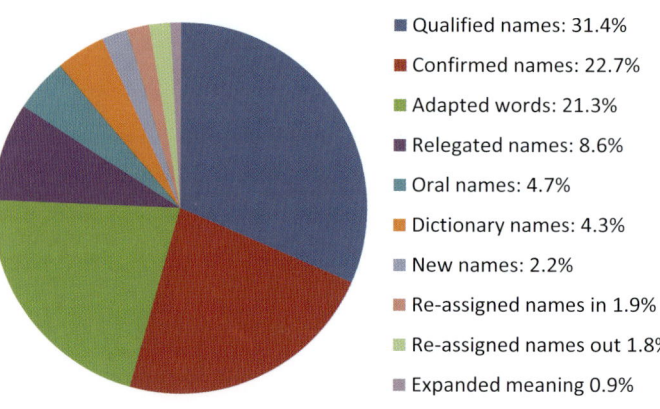

- Qualified names: 31.4%
- Confirmed names: 22.7%
- Adapted words: 21.3%
- Relegated names: 8.6%
- Oral names: 4.7%
- Dictionary names: 4.3%
- New names: 2.2%
- Re-assigned names in 1.9%
- Re-assigned names out 1.8%
- Expanded meaning 0.9%

4. Conclusion

Native and traditional wisdom has historically provided a large body of knowledge about the environment embedded in indigenous culture. The nature of indigenous knowledge, usually handed down orally, has assisted communities in adapting to and surviving in a dynamic environment. Its application in biodiversity conservation, research, monitoring and management has however, not been fully explored. Identification of an organism or organisms is fundamental to reporting, research, monitoring, and even coining of names. If one cannot identify a plant, animal or bird by a unique individual name, then one cannot share or record vital information about the species. Biodiversity conservation efforts require an integrated two-way communication approach in the involvement of people as major stakeholders.

Avian nomenclature in the African languages of southern Africa has not previously received the attention it deserves. It is only now that some attention is being given to this important matter. It is necessary to understand the linguistics of naming systems particularly for a popular taxon such as birds, so as to unravel the potentials of indigenous people in bird conservation, research, and population monitoring.

Names are dynamic and subject to a number of external forces that instigate name changes. There will no doubt be Zulu-speaking people who will not agree with the name chosen as the standard. This is inevitable given that so many different names for one species are regional in nature, or when respondents from different age groups are using names popular fifty years ago but not so commonly used nowadays.

Achievement of goals

In the Preamble at the beginning of this Introduction, it was stated that the aim of this project was to:

1. Identify and record existing vernacular Zulu names for all birds found in the KwaZulu-Natal region:

We will probably never know whether we have recorded every single name currently used for birds in the various regions of the province. It seems likely there will be a few existing names which have escaped us.

2. Select one commonly used name for a species when two or more names existed:

For the most part this was done. However, when in some cases two names were identified as being equally well-known and widely distributed throughout KwaZulu-Natal, both became the standard names.

3. Distinguish between various species having the same name:

This was achieved for all 550 species of birds listed.

4. Reassign names where necessary:

This was done for only a small percentage (1.9%) of the names.

5. Coin and create new names where none had existed before:

This was achieved either by adapting and extending previously recorded names or by coining a completely new word.

CONSERVATION

5. Conservation status of birds in KwaZulu-Natal

In the most recent assessment of the conservation status of birds in the South African region (that includes South Africa, Lesotho and Swaziland) Taylor et al. (2015) concluded that in comparing the previous 2000 assessment (127 species) with the 2015 assessment (132 species) threats affecting the avifauna had not abated adequately over this time. Rather the changes found had moved previously listed species into higher endangered threat categories, therefore more species were listed in 2015 in the Endangered and Vulnerable groups and remarkably few species left the list of threatened species during this period of time. Indeed it is most troubling that the efforts to mitigate the severity of or eliminate threats and their impact, had not reversed the trends for most species leading one to the conclusion that extinction for many of our bird species is looming.

Critically Endangered Species in KwaZulu-Natal

Of South Africa's 13 regionally critically endangered bird species eight are found in KwaZulu-Natal, namely **ubhamukwe** (Wattled Crane), **indlanyokempunga** (Southern Banded Snake Eagle), **ubhavuzile omhlophe** (White-winged Flufftail), **inkonjanesibhakabhaka** (Blue Swallow), **ukhozilwentshebe** (Bearded Vulture), **inqelincane** (Hooded Vulture), **inqe lehlanze** (White-backed Vulture), and **ukhandelimhlophe** (White-headed Vulture) (Taylor et al. 2015). All of these species are considered to be facing an extremely high risk of extinction in the wild. This is a 400% increase in the number of critically endangered birds in the country since the previous assessment that was conducted in 2000. In spite of a lot of research, conservation work and money they remain on the verge of extinction in the country as a whole, for example both Wattled Crane and Blue Swallow have been on this list for more than 19 years. Five vulture species found in KwaZulu-Natal (that include the four species listed above plus the regionally endangered **indlangamandla** (Lappet-faced Vulture) teeter on the edge of extinction in the province with the White-headed Vulture in all likelihood now extinct in 2019 and the Lappet-faced and Hooded Vultures close behind. Certainly much more must be done by Ezemvelo KZN Wildlife (the Provincial mandated conservation agency), BirdLife South Africa, the Endangered Wildlife Trust and other civil conservation organizations to minimize the threats that are driving all these species towards their extinction.

Endangered Species in KwaZulu-Natal

There are 38 listed regionally endangered bird species all of which face a very high risk of extinction in the wild. Of these some 20 species occur in KwaZulu-Natal excluding listed endangered seabirds (Taylor et al. 2015). These are:

unongoyana (Green Barbet), **ingqungqulu** (Bateleur), **ungoqo** (Black-rumped Buttonquail), **unohemu** (Grey Crowned Crane), **inkosiyezinkozi** (Martial Eagle), **ukhozolunsundu** (Tawny Eagle), **insingizi** (Southern Ground Hornbill), **umamhlangenonsundu** (African Marsh Harrier), **umamhlangeni omnyama** (Black Harrier), **umahlawithilulwane** (Bat Hawk), **unonkalankala** (Mangrove Kingfisher), **isikhovanhlanzi** (Pel's Fishing Owl), **isikhwenene** (Cape Parrot), **ijubelintamemhlophe** (Eastern Bronze-naped Pigeon), **umadolabomvu** (Saddle-billed Stork), **unomlomophuzi** (Yellow-billed Stork), **umunswi wehlathi** (Spotted Ground Thrush), **idlanga lentaba** (Cape Vulture), **indlangamandla** (Lappet-faced Vulture).

Vulnerable Species in KwaZulu-Natal

There are 23 vulnerable bird species in KwaZulu-Natal (33 species in the South African region that includes several seabirds) that are considered to be facing a high risk of extinction in the wild. This group has a large number of the larger-bodied birds including raptors, grassland and wetland specialists. Therefore they bring to our attention that the grassland biome, wetlands, rivers and estuaries are inadequately protected and managed. Destructive, detrimental and unsustainable land use practices such as overgrazing, transformation through clearing of vegetation, uncontrolled invasion by aggressive alien plant species, excessive veld burning, draining as well as direct and indirect disposal of pollutants into rivers and other water bodies, are activities that all have long-term adverse impacts on the avifauna of the region. Such activities are resulting in the population decline of our bird species, in particular, driving many of them into an extinction vortex. To address and reverse this trend requires that land-owners in KwaZulu-Natal and elsewhere protect key habitat patches and that government authorities ensure their compliance with the environmental laws and regulations, and severely punish repeat offenders.

CONSERVATION

The vulnerable species are: **unosigqokomnyama** (Bush Blackcap), **umasikulufu** (African Broadbill), **iseme** (Denham's Bustard), **ukhoziqholwane** (Crowned Eagle), **ukhozolumnyama** (Verreaux's Eagle), **uklebemawa** (Lanner Falcon), **igwedlamanzi** (African Finfoot), **ubhavuzilomidwayidwa** (Striped Flufftail), **isicibamanzi** (Cape Gannet), **ivevenyane** (African Pygmy Goose), **umacuthobomvu** (White-backed Night Heron), **umxwagele** (Southern Bald Ibis), **unondwayizomncane** (Lesser Jacana), **inkakulo** (White-bellied Bustard), **umhlohlongwane** (Swamp Nightjar), **isikhova sotshani** (African Grass Owl), **ivubelikhulu** (Great White Pelican), **ivubelincane** (Pink-backed Pelican), **umngcelwane** (Short-tailed Pipit), **ungcelekeshephuzi** (Yellow-breasted Pipit), **intinginono** (Secretarybird), **unowanga** (Black Stork), and **ubaklakliyo** (Caspian Tern).

Near Threatened Species

This category includes those bird species of concern, usually because data is insufficient but populations may be considered to be either small, fragmented, declining or their distribution range is becoming increasingly restricted. Importantly, current threats are seen not to be that severe to qualify for the species to be placed in the vulnerable category (Taylor *et al.* 2015).

For KwaZulu-Natal the following near threatened species have been listed: **umalaleni** (Lemon-breasted Canary), **indwe** (Blue Crane), **idadelikhandamnyama** (Maccoa Duck), **uhlebonyawobomvu** (Red-footed Falcon), **ukholwasa omkhulu** (Greater Flamingo), **ukholwasomncane** (Lesser Flamingo), **isixula** (Half-collared Kingfisher), **inqomfebomvu** (Rosy-throated Longclaw), **umakhwaneni** (Greater Painted Snipe), **umncgelu wasematsheni** (African Rock Pipit), **umngcelu** (Mountain Pipit), **ifefeluhlaza** (European Roller), **inqelendlovu** (Marabou Stork), **inswinswi** (Orange Ground Thrush), **isigqobhamithi saseningizimu** (Knysna Woodpecker).

Extinct bird species in KwaZulu-Natal

There are three species of birds that have become extinct in KwaZulu-Natal. They are: Burchell's Courser, Rudd's Lark, and African Skimmer, and possibly the Egyptian Vulture given its historical significance in Zulu culture (Koopman *et al.* 2020).

South African endemic and near-endemic species occurring in KZN

In addition to the many South African endangered bird species that occur in KwaZulu-Natal, the province has some 54% of the region's endemic species. These are: **uklebe lwehlathi** (Forest Buzzard), **inhlandlokazi** (Jackal Buzzard), **umbhalane wehlathi** (Forest Canary), **inkuletsheni** (Buff-streaked Chat), **isaqola** (Fiscal Flycatcher), **ithendelelimlotha** (Grey-winged Francolin), **unontshiloza** (Cape Grassbird), **umbukwane** (Blue Korhaan), **unonzwili** (Eastern Long-billed Lark), **ujenga wokhahlamba** (Drakensberg Prinia), **ugaganonsundu** (Brown Scrub Robin), **umananda** (Chorister Robin-Chat), **unogxumetsheni** (Drakensberg Rockjumper), **ingwangwa** (Pied Starling), **incuncu** (Greater Double-collared Sunbird), **incuncwana** (Southern Double-collared Sunbird), **inkonjane yamawa** (South African Cliff Swallow), **ushowe** (Southern Tchagra), **isihlalamatsheni** (Cape Rock Thrush), **ikhwelematsheni** (Sentinel Rock Thrush), **igwalagwaleliluhlaza** (Knysna Turaco), **ubusukuswane** (Swee Waxbill), **ihlokohlokelikhulu** (Cape Weaver), **umehlwanoluhlaza** (Cape White-eye), and **umnqangqandolo** (Ground Woodpecker).

Together with the endemic and the near-endemic bird species that KwaZulu-Natal shares with Mozambique (such as **umankole** (Rudd's Apalis), **umnqube wogu** (Woodwards' Batis), **umagumejana** (Pink-throated Twinspot), **igwalagwala logu** (Livingstone's Turaco), **incwincwi yaseTembu** (Plain-backed Sunbird), the province is an important destination for both the domestic and international avifauna tourism industry.

Fostering care and an appreciation of our bird heritage

The bird species listed above are the prime targets for bird watchers and the growing avitourism industry in KwaZulu-Natal. It is this rich heritage of bird life with about 78% of South Africa's recorded species that is the foundation for this industry. The industry thrives on world-class protected areas and up-market facilities. It is supported by tour companies specializing in avitourism that employ trained professional bird guides.

CONSERVATION

KwaZulu-Natal Important Birding Areas

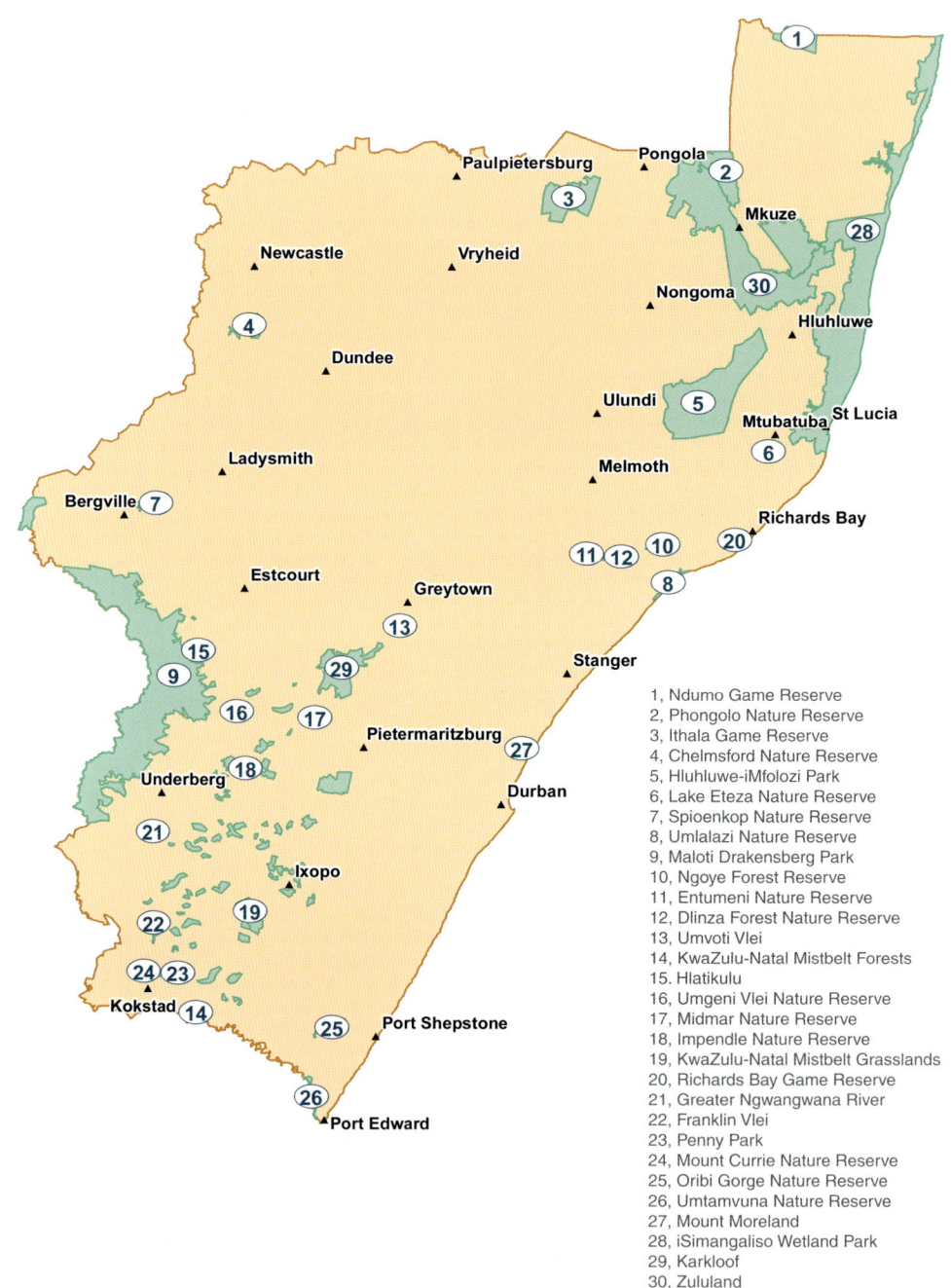

1, Ndumo Game Reserve
2, Phongolo Nature Reserve
3, Ithala Game Reserve
4, Chelmsford Nature Reserve
5, Hluhluwe-iMfolozi Park
6, Lake Eteza Nature Reserve
7, Spioenkop Nature Reserve
8, Umlalazi Nature Reserve
9, Maloti Drakensberg Park
10, Ngoye Forest Reserve
11, Entumeni Nature Reserve
12, Dlinza Forest Nature Reserve
13, Umvoti Vlei
14, KwaZulu-Natal Mistbelt Forests
15, Hlatikulu
16, Umgeni Vlei Nature Reserve
17, Midmar Nature Reserve
18, Impendle Nature Reserve
19, KwaZulu-Natal Mistbelt Grasslands
20, Richards Bay Game Reserve
21, Greater Ngwangwana River
22, Franklin Vlei
23, Penny Park
24, Mount Currie Nature Reserve
25, Oribi Gorge Nature Reserve
26, Umtamvuna Nature Reserve
27, Mount Moreland
28, iSimangaliso Wetland Park
29, Karkloof
30, Zululand

CONSERVATION

Within the landscape of KwaZulu-Natal are the mountain, grassland, forest, woodland, wetland and coastal areas that provide the natural habitats for this large variety of birds. Avitourism hotspots include the Maloti Drakensberg Park World Heritage Site, iSimangaliso Wetland Park World Heritage Site (particularly certain sections, e.g. uMkhuze, oZabeni, Eastern Shores, St Lucia Estuary and Kosi system), uMlalazi Estuary, oNgoye and Dukuduku Forests, Tembe Elephant Park, Ndumo Game Reserve, Hluhluwe-iMfolozi Game Reserve, Ithala Game Reserve, the Franklin, Blood River, uMvoti, eNsikeni and uMngeni marshes, the Krantzkloof, Oribi, uMtamvuna gorges, and the grasslands of the Midlands.

Although many of these areas have formal conservation protection, some 60% of listed Important Bird and Biodiversity Areas of South Africa (IBA) have no legal protection and are found on private and state land. Importantly there are 29 IBAs listed for KwaZulu-Natal having a combined total area of 731 120 ha (Marnewick *et al.* 2015). Although the protected area network is extensive, it has been shown to be inadequate for protecting and sustaining the bird life of our country and province.

Birds face many threats in our modern world. These threats are increasing (Taylor *et al.* 2015) as the human population grows which is expected to exceed 10 billion people by the end of this century (Tilman *et al.* 2017). Given the increasing demand for space and croplands, human needs and their impacts on the environment are endangering birds and their habitats leading to an acceleration of the rate of extinctions. Modification and fragmentation of natural areas are the most frequent and direct drivers of biodiversity loss. Populations of species therefore become increasingly confined into smaller and smaller isolated patches which can no longer support their survival. Examples of human activities and drivers of such loss are: urbanization, agriculture, mining, development of transport corridors, invasive species, and water pollution.

In KwaZulu-Natal much of the land area of the province has been modified (Jewitt *et al.* 2015). She found that the drivers of this habitat loss were agriculture, plantation forestry, urbanization and industrial development, dams and mines.

Also significant is the use of insecticides (and other biocides) especially when applied to croplands as an aerial spray. This has resulted in a decline in insect populations, thus decreasing the food for insectivorous bird species. Such chemicals may also enter water bodies and affect aquatic organisms that are fed on by birds.

Indiscriminate poisoning has had a devastating effect on the vulture population particularly in northern KwaZulu-Natal (as well as elsewhere) and is responsible for the catastrophic decline to near extinction of both the White-headed and Lappet-faced Vultures and is driving the White-backed Vulture rapidly towards the same fate. Poachers of wild animals poison carcasses to prevent soaring vultures from giving away the location of a dead animal to law enforcement personnel. In addition, there is a demand and trade in vulture parts for traditional beliefs: as these birds can see very far, it is believed that they locate their food with clairvoyant ability, and that they can see into the future. A purchaser of vulture muti believes in the myth that he would be able to see into the future and thus be able to gamble successfully in any game of chance (Koopman 2019). The result of these beliefs threatens vultures' survival since demand for birds and their body parts for various purposes may increase with human expansion (Turner & Koopman, 2018).

Overhead power-lines and fences constitute a major threat to large terrestrial and wetland birds. Flamingos, cranes, bustards, also Secretarybird and Ground Hornbill, collide directly into these structures. Electrocution of large birds such as vultures and eagles results if the pylons that support the live conductors allow for places where the birds can perch hence enabling the bird's open wings to make contact with the live wires.

To counter such threats, all-embracing environmental legislation that controls and regulates proposed new developments is in place. But assessment of a proposal for development relies on concerned citizens and organizations submitting their concerns, usually by engaging with the Environmental Impact Assessment process. For this to happen effectively people need to be aware of and be able to express their concerns. Fortunately there are several organizations that play important roles in promoting conservation awareness.

The private sector, for example, is very active in addressing threats to biodiversity. Several organizations promote the conservation of birds with BirdLife South Africa leading the way. This non-governmental organization strives to conserve birds and their habitats. By encouraging people to enjoy and value nature through several outreach programmes and events, the organization inculcates awareness, care and an appreciation of the rich diversity of birds in southern Africa among its members and associates.

CONSERVATION

Events include bird fairs and the Birding Big Day competitions. The magazine *African BirdLife* and their monthly e-newsletter sent to some 10 000 people keep members and subscribers informed. Importantly, Birdlife SA has developed a comprehensive conservation strategy with programmes addressing individual matters of concern such as the country's threatened birds, and completing the Important Bird Areas (IBA) strategy in 2018. They are able to monitor and assess changes in land cover and climate through their Conservation Modeling Project. An important facility is the Wakkerstroom Education and Tourism Centre.

The Field Guide Association of South Africa (FGASA) offers qualifications in bird guiding, certifying learners as either local, regional, national or Special Knowledge and Skills (SKS). The course consists of a balanced proportion of theoretical and practical activities. An aspirant bird guide needs to have attained a FGASA Field Guide, or an Advanced Field Guide, or the FGASA Specialist Field Guide qualification after having undergone training and assessment. Details of these qualifications can be found on the FGASA website (https://www.fgasa.co.za). The requirements of the training courses are demanding for professional certification. The most demanding is the SKS (Birding) Qualification where a candidate must be competent in three areas: birding theoretical, bird identification, and bird guiding. The candidate must then pass a photograph and sound assessment for a chosen region. This is followed by a theory examination in eleven subject areas such as bird classification and conservation, bird behaviour, breeding, feeding, migration and navigation.

Qualified professional bird guides have been placed in internship positions at appropriate locations such as game lodges, and some of these have become freelance and now operate their own businesses. Use was made of 18 experienced qualified bird guides all located in KwaZulu-Natal who participated in a series of six workshops that resulted in providing the information for this book, as well as two other books on Zulu bird names published by the University of KwaZulu-Natal Press, namely: *Zulu Bird Names and Bird Lore* (2019) and *Amagama Ezinyoni: Zulu Names of Birds* (2020). Some of these guides have embarked on environmental education programmes as part of their business. They engage with both teachers and learners bringing the conservation message to schools where learners become involved in various practical activities that reinforce what has been learnt in the class room. Teachers attend workshops during the year that focus on how to implement Biodiversity Stewardship thereby ensuring cross-curriculum learning.

Other initiatives include Birdlife South Africa-affiliated bird clubs in Durban, Zululand, Natal Midlands and the Natal South Coast, that arrange bird-watching outings and presentations on birds often by experts on various subjects. The Southern African Bird Atlas Project (SABAP 2) based at the University of Cape Town and initiated in 2007, has resulted in a growing interest and involvement by both young and old in this mapping of bird distribution. The Education Department of the Durban Natural Science Museum conducts environmental outreach programmes on a day-to-day basis. In addition, their 'Go-Wild' Mobile museum visits schools and communities in the eThekwini area. The vehicle is staffed by trained volunteers and has specimens of birds and other animals as well as educational material. The programmes offered are aligned to the school curriculum and are grade-specific from Grade R through to Grade 12. These ensure a greater understanding of class work.

Although there are these functional education programmes operating in KwaZulu-Natal, what has become apparent is that there is a dearth of literature resources available in the home language of the majority of learners participating in these programmes. The potential and opportunities provided by birds as subject matter for readers is considerable. There is no reason why Zulu traditional bird lore, and the explanation of the Zulu bird names as they relate to this heritage, should not become a staple content of readers written for children in Zulu. Certainly the life of birds, their diversity, biology and different behaviours, their cultural significance and importance, provide a vast amount of material for school text books written in Zulu as examples of a wide range of theories and natural science concepts.

Literature on birds in Zulu is required for ornithological teaching at tertiary level at universities, and Zulu bird names would become essential for lectures, class notes and other written teaching material. The Zoology Department, University of KwaZulu-Natal, (and its predecessor the University of Natal) has a widely acknowledged reputation in producing top-class graduates in ornithology many of whom have become leaders in their field of research. With increasing numbers of first language Zulu students undertaking post-graduate ornithological studies, the need for avifauna literature in Zulu will become ever more pressing.

BIRDS STIMULATE IMAGINATION AND INSPIRE CREATIVITY

Calabash by Adrian Koopman. Remaining birds in wood, seeds, beads and wire, and embroidered, all by various artists.

OSTRICH, GUINEAFOWL, PEAFOWL AND FRANCOLIN

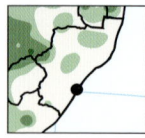

intshe Common Ostrich *Struthio camelus*
While there is no doubt about this Zulu name, there is debate about whether or not this bird occurred historically in KwaZulu-Natal. There is certainly strong evidence that ostrich feathers were used regularly as military insignia and regalia during the nineteenth century, and it would appear that the Zulus traded with the Boers further north, exchanging cattle for feathers.

GUINEAFOWL GENERIC NAME: impangele

impangelejwayelekile Helmeted Guineafowl *Numida meleagris*
The name **impangele** is well-known. Here *ejwayekile* (meaning 'common') is added to distinguish it from the Crested Guineafowl. A well-known Zulu proverb is *impangel' enhle ikhal' igijima* (lit. 'a good guineafowl calls as it runs') meaning 'Make the best of a bad job'.

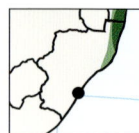

impangele yehlathi Crested Guineafowl *Guttera pucherani*
The use of **impangele** with the extension *yehlathi* 'of the forest' is well-known.

upigogo Indian Peafowl *Pavo cristatus*
The Zulu name **upigogo** has been adapted from the English name *peacock*, and so has nothing to do with the similar-sounding Zulu question "*uphi gogo*" ('Where is grandmother?').

FRANCOLIN AND SPURFOWL GENERIC NAMES: intendele and isikhwehle

inswempe Coqui Francolin *Peliperdix coqui*
The name **inswempe** is imitative of the characteristic call of this bird, and has also given rise to the Afrikaans name *swempie* for this bird.

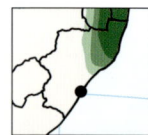

isikhwehlesiqhova Crested Francolin *Dendroperdix sephaena*
The word **isikhwehle** is one of the generic terms for francolins and spurfowl, and the qualification [i]*siqhova* 'crest' has been added to it for the Crested Francolin.

impangelejwayelekile Helmeted Guineafowl flock

intshe Common Ostrich

impangelejwayelekile Helmeted Guineafowl

impangele yehlathi Crested Guineafowl

upigogo Indian Peafowl ♂

inswempe Coqui Francolin ♂

isikhwehlesiqhova Crested Francolin

29

SPURFOWL AND FRANCOLIN

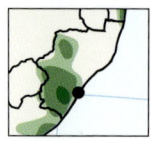

isikhwehlesimqalabomvu Red-necked Spurfowl *Pternistis afer*

The generic name **isikhwehle** has been extended with *esimqalabomvu* 'red-necked'.

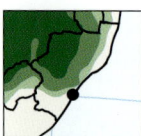

inkwali Swainson's Spurfowl *Pternistis swainsonii*

The word **inkwali** is clearly a very old Zulu word, from the Ur-Bantu root **kwale** meaning 'francolin'. Today this name is used specifically for Swainson's Spurfowl. A well-known Zulu proverb is *akukh' inkwal' ephandel' enye* (lit. 'there is no francolin that scratches for another'), meaning 'everybody should look out for himself'.

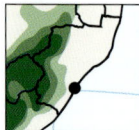

ithendelelimlotha Grey-winged Francolin *Scleroptila afra*

The generic name **ithendele** also occurs regularly in the form **intendele**. The form **ithendele** has been extended here with **elimlotha** 'which is ash-coloured'.

There are two proverbs in Zulu which relate to the hunting of this bird. One is *intendel' esuka muva ikholwa yizagila* (lit. 'the francolin that rises last gets struck by the throwing stick', meaning 'when danger threatens, get away quickly; don't wait around'). The other is *intendel' iw' enkundleni* (lit. 'the francolin has fallen into the yard'), meaning 'we have been very lucky in receiving an unexpected windfall'.

A third proverb relates to planting seed: a bulb of the *isidwa* gladiolus (*Gladiolus ludwigii*) is placed in the bag of seed when planting to ensure a good harvest. It may get thrown out by accident and pecked at by mistake by a francolin eating all the planted seed. The proverb is *ithendele ibindwa yisidwa* ('the francolin chokes on the *isidwa* bulb'), meaning 'a person rendered speechless when caught red-handed'. All these three proverbs relate to the Red-winged Francolin as well.

ithendelelibomvu Red-winged Francolin *Scleroptila levaillantii*

The francolin generic **ithendele** has been extended here with *elibomvu* 'which is red'.

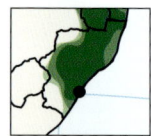

isikhwehlesimqhalomhlophe Shelley's Francolin *Scleroptila shelleyi*

The francolin generic **isikhwehle** has been extended with *esimqhalomhlophe*, lit. 'white-throated'.

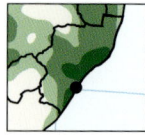

unomemeza Natal Spurfowl *Pternistis natalensis*

This name is based on the incessant calling of this bird. The name **unomemeza** means 'the one that calls continuously', and it is derived from the verb *memeza* 'to call'.

isikhwehlesimqalabomvu Red-necked Spurfowl

inkwali Swainson's Spurfowl

ithendelelimlotha Grey-winged Francolin

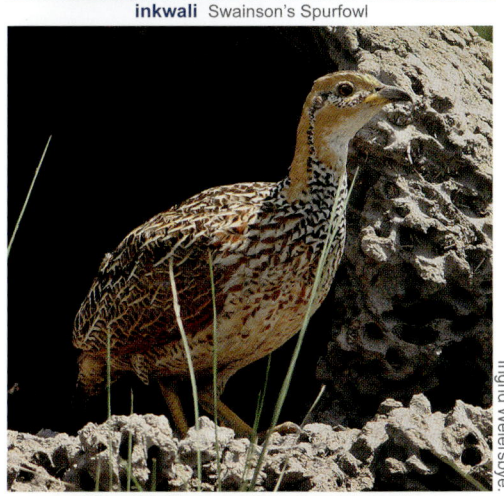
ithendelelibomvu Red-winged Francolin

isikhwehlesimqhalomhlophe Shelley's Francolin

unomemeza Natal Spurfowl

QUAIL AND BUTTONQUAIL

QUAIL GENERIC NAME: isagwaca . This long-established name is derived from the verb *gwaca* 'lie flat', 'lie low in hiding', prefixed with –*sa*– 'something like'.

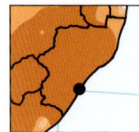

isagwacesijwayelekile Common Quail *Coturnix coturnix*

There are two forms of this name – **isagwaca** and **isigwaca** – for a quail, **isagwaca** being the more common form. Both forms are derived from the Zulu verb *gwaca* ('lie flat, lie low, lie in hiding'). The extension **esijwayelekile** 'which is common' has been added to distinguish this bird from the Harlequin Quail.

A Zulu proverb says *isagwaca silindel' induku* (lit. 'the quail waits for the hunting stick'), applied to any stupid person who waits for trouble to overtake him and doesn't try to do anything to prevent it.

isagwacesibomvu Harlequin Quail *Coturnix delegorguei*

The generic **isagwaca** has been extended with *esibomvu* 'which is red'.

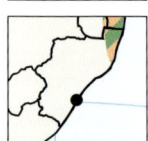

isagwacesibhakabhaka Blue Quail *Excalfactoria adansonii*

The generic name **isagwaca** has been extended with *esibhakabhaka* 'which is blue like the sky'.

ungolwane Common Buttonquail *Turnix sylvaticus*

There is no underlying meaning of the name **ungolwane**.

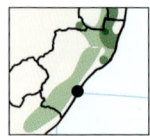

ungoqo Black-rumped Buttonquail *Turnix nanus*

There is no underlying **meaning** of the name **ungoqo**.

ungoqo Black-rumped Buttonquail ♀

isagwacesijwayelekile Common Quail

isagwacesibomvu Harlequin Quail ♂

isagwacesibomvu Harlequin Quail ♀

isagwacesibhakabhaka Blue Quail ♂

ungolwane Common Buttonquail ♀

ungoqo Black-rumped Buttonquail ♀

33

WHISTLING DUCK, DUCK AND GOOSE

DUCK GENERIC NAME: idada. All species of duck have previously been called '**idada**'. For some of the species below, this generic term has been extended with one or the other qualifying phrase; for others totally new names have been coined.

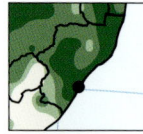

inzwinzwinzwi White-faced Whistling Duck *Dendrocygna viduata*
This is an adaptation of the onomatopoeia-based name **inzwinzwi** coined for the White-backed Duck (see below).

inzwinzwebomvu Fulvous Whistling Duck *Dendrocygna bicolor*
The coined name **inzwinzwi** has been extended with *ebomvu* 'which is red' to give a name which refers to both the whistling sound of this duck as well as its dominant red colour.

inzwinzwi White-backed Duck *Thalassornis leuconotus*
The whistling sound made by this species is seen to be more salient than the white of its back, hence the imitative name **inzwinzwi**.

ivevenyane African Pygmy Goose *Nettapus auritus*
The name **ivevenyane** is a well-known name. It is possibly derived from *uveve* 'horn trumpet', 'horn, wood or reed whistle', suffixed by – *nyane*, commonly found in the names of species of living things. If so, then this is a call-based name.

ihhoye Spur-winged Goose *Plectropterus gambensis*
The name **ihhoye** is widely-used today, and has no obvious underlying meaning.

ilongwe Egyptian Goose *Alopochen aegyptiaca*
This is a well-known name for the bird.

inzwinzwi White-backed Duck and chicks

inzwinzwinzwi White-faced Whistling Duck

inzwinzwebomvu Fulvous Whistling Duck

inzwinzwi White-backed Duck

ivevenyane African Pygmy Goose ♂

ihhoye Spur-winged Goose

ilongwe Egyptian Goose

DUCK, SHELDUCK AND TEAL

unosimila Knob-billed Duck *Sarkidiornis melanotos*
The name **unosimila** is based on the word *isimila* 'growth', and refers to the knob-like growth on the beak of this bird.

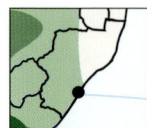

idadelibomvu South African Shelduck *Tadorna cana*
The generic name **idada** has been extended with *elibomvu* 'which is red'.

unosikhutha Cape Teal *Anas capensis*
The name **unosikhutha** is based on the noun *isikhutha* 'mould, mildew', and refers to the pale mottled colour of the bird, reminding one of mouldy bread.

idadelimlomophuzi Yellow-billed Duck *Anas undulata*
A straightforward extension of the generic **idada** with *elimlomophuzi* 'yellow-billed', a compound of *umlomo* 'mouth, beak' and *phuzi* 'yellow'.

idada laseYurobhu Mallard *Anas platyrhynchos*
This duck is an introduced species from Europe and, fittingly, the generic name **idada** has been extended with *laseYurobhu* 'from Europe'.

idadelimnyama African Black Duck *Anas sparsa*
The generic **idada** has been extended with *elimnyama* 'which is black'.

unosimila Knob-billed Duck ♀

idadalibomvu South African Shelduck ♀

unosimila Knob-billed Duck ♂

idadelibomvu South African Shelduck ♂

unosikhutha Cape Teal

idadelimlomophuzi Yellow-billed Duck

idada laseYurobhu Mallard ♂

idadelikhandamnyama African Black Duck

SHOVELER, TEAL, POCHARD AND DUCK

unofosholo Cape Shoveler *Anas smithii*
The Zulu name **unofosholo** comes from the ordinary noun *ifosholo*, which in turn comes from the English word 'shovel'.

idadelimlomobomvu Red-billed Teal *Anas erythrorhyncha*
The generic **idada** has been extended here with **elimlomobomvu** 'red-billed'.

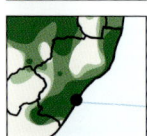

idadelincane Hottentot Teal *Anas hottenttota*
The generic **idada** has been extended with *elincane* 'which is small'.

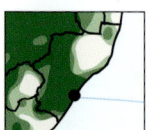

isankawu Southern Pochard *Netta erythrophthalma*
Because the call of this bird is seen to be similar to the alarm call of a vervet monkey (*inkawu*), this word has been prefixed with *–sa–* 'something like' to form **isankawu** 'something like a vervet monkey'.

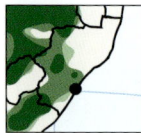

idadelikhandamnyama Maccoa Duck *Oxyura maccaoa*
The generic name **idada** has been extended here with **elikhandamnyama** 'which is black-headed'.

unofosholo Cape Shoveler ♀ and ♂

unofosholo Cape Shoveler ♂

idadelimlomobomvu Red-billed Teal

idadelincane Hottentot Teal

isankawu Southern Pochard ♀ and ♂

idadelikhandamnyama Maccoa Duck ♀

idadelikhandamnyama Maccoa Duck ♂

PENGUIN, ALBATROSS, SHEARWATER, PETREL, PRION AND GREBE

PENGUIN GENERIC NAME: inguza. This covers all penguins.

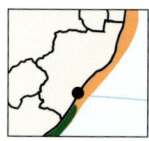

inguzambongolo African Penguin *Spheniscus demersus*
The species-specific name for this bird follows the old English name 'Jackass Penguin' by compounding **inguza** with *imbongolo* 'donkey', 'jackass', and the resultant **inguzambongolo** is thus a reference to the braying call of this bird.

PELAGIC BIRDS GENERIC NAMES: The following were created for albatrosses, shearwaters, petrels, storm petrels and prions: No individual Zulu names were given to any individual species of these marine birds.

unozulane Albatross
The generic name **unozulane** is based on the verb *zula* 'to wander about'.

isikamanzi Shearwater
(a compound of *sika* (cut) and *amanzi* (water); i.e. 'what cuts the water')

umantantolwandle Petrel
(a compound based on *ntanta* 'float above' and *ulwandle* 'sea', i.e. 'what floats just above the ocean')

unohhalulwandle Storm Petrel
(a compound of *hhala* 'rake' + *ulwandle* 'sea', a reference to the way the feet of this bird trail in the water as it flies above the surface of the ocean)

impungayolwandle Prion
(a compound of *mpunga* 'grey' and *yolwandle* 'of the sea', i.e. 'little grey [bird] of the ocean')

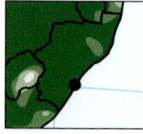

imvukwane Little Grebe [Dabchick] *Tachibaptus ruficollis*
This name is likely derived from the noun *uvuko* 'a rising up' to which the suffix *–ane* has been added. This would be a reference to the bird bobbing up again after diving.

imvukwane Little Grebe [Dabchick] with frog prey

inguzambongolo African Penguin

Yellow-nosed Albatross

Flesh-footed Shearwater

White-chinned Petrel

Wilson's Storm Petrel

Slender-billed Prion

FLAMINGO, PELICAN, STORK AND OPENBILL

FLAMINGO GENERIC NAME: ukholwase

ukholwasomkhulu Greater Flamingo *Phoenicopterus roseus*
To distinguish this bird from the Lesser Flamingo, the generic **ukholwase** was extended with *omkhulu* 'large'.

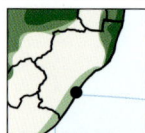

ukholwasomncane Lesser Flamingo *Phoeniconaias minor*
The generic **ukholwase** has been extended with *omncane* 'small' to distinguish this bird from the Greater Flamingo.

PELICAN GENERIC NAME: ivuba

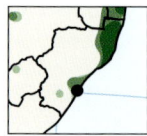

ivubelincane Pink-backed Pelican *Pelecanus rufescens*
The generic **ivuba** has been extended with *elincane* 'small' to distinguish this bird from the Great White Pelican.

ivubelikhulu, ikhungula Great White Pelican *Pelecanus onocrotalus*
The generic **ivuba** has been extended with *elikhulu* 'large' to distinguish this bird from the Pink-backed Pelican. The name **ikhungula** has not been previously recorded, but is otherwise well-known.

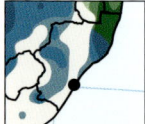

unomlomophuzi Yellow-billed Stork *Mycteria ibis*
This name, a compound of *–no–* with *umlomo* 'mouth, beak' and *phuzi* 'yellow', follows the English name in meaning.

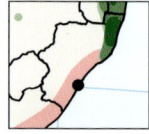

isigqobhamnenke African Openbill *Anastomus lamelligerus*
This name is a compound of *gqobha* 'peck at' and *umnenke* 'snail', and refers to a major food item of this bird.

ukholwasomkhulu Greater Flamingo feeding

ukholwasomncane Lesser Flamingo feeding

ukholwasomkhulu Greater Flamingo

ukholwasomncane Lesser Flamingo

ivubelincane Pink-backed Pelican

ivubelikhulu Great White Pelican

unomlomophuzi Yellow-billed Stork

isigqobhamnenke African Openbill

STORK

unowanga Black Stork *Ciconia nigra*
One of the 'extra' or 'unused' names for the White Stork was transferred to this species. It has no obvious underlying meaning.

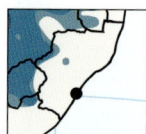
umahlombamhlophe Abdim's Stork *Ciconia abdimii*
This descriptive name is a compound of *amahlombe* 'shoulders'; and *amhlophe* 'which are white'.

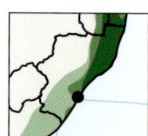
isithandamanzi Woolly-necked Stork *Ciconia episcopus*
This well-known name for this bird is a compound of *thanda* 'like' and *amanzi* 'water'.

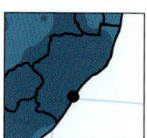
unogolantethe White Stork *Ciconia ciconia*
The name means 'the one that catches locusts', and it is interesting to note that an earlier colonial name for this bird was Locust Bird.

[Alternative, but less well-known names are **unowanga**, **unoyenge**, and **ingababa**, none of which have an obvious underlying meaning.]

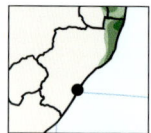
umadolabomvu Saddle-billed Stork *Ephippiorhynchus senegalensis*
This descriptive name is a compound of *amadolo* 'knees' and *abomvu* 'which are red'.

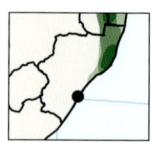
inqelendlovu Marabou Stork *Leptoptilos crumenifer*
The name **inqelendlovu** is a compound of *inqe* 'vulture' and *lendlovu* 'of the elephant'. The bird's head and neck are distinctly vulture-like and it is 'elephant-sized' in comparison to the true vultures.

isithandamanzi Woolly-necked Stork with fishing line on foot

unowanga Black Stork

umahlombamhlophe Abdim's Stork

isithandamanzi Woolly-necked Stork

unogolantethe White Stork

umadolabomvu Saddle-billed Stork

inqelendlovu Marabou Stork

IBIS AND SPOONBILL

umxwagele Southern Bald Ibis *Geronticus calvus*

The name **umxwagele** is well-known and long-established. It has no obvious underlying meaning.

inkankanelunga Sacred Ibis *Threskiornis aethiopicus*

The name **inkankanelunga** is a compound of **inkankane** 'Hadada Ibis' and *ilunga*, a name primarily indicating a black and white cow, but used here (and in other Zulu words) as the equivalent of 'pied'. Effectively, then, the name could be translated as 'Pied Hadada'.

inkankane, ihhahane Hadada Ibis *Bostrychia hagedash*

Both these names, imitating the call of the bird, are well-known. In some areas the bird is also known as **unongqanga**.

umacibudaka Glossy Ibis *Plegadis falcinellus*

This name is based on *–ma–* 'characteristic', the verb *ciba* (used here with the meaning 'do at one's pleasure') and *udaka* 'mud', thus 'the bird that characteristically takes pleasure in mud'.

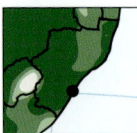

isixulamasele African Spoonbill *Platalea alba*

This name is based on the verb *xula* 'seize up', 'grab at', 'catch unawares', and the plural noun *amasele* 'frogs'. The bird is also widely known as **inkenkane**.

umxwagele Southern Bald Ibis showing how it uses its long bill to catch food

umxwagele Southern Bald Ibis

umxwagele Southern Bald Ibis (juv)

inkankanelunga Sacred Ibis

inkankane Hadada Ibis

umacibudaka Glossy Ibis

isixulamasele African Spoonbill

BITTERN, HAMERKOP AND NIGHT HERON

BITTERN/HERON GENERIC NAMES: umabhu and umacutha. The generic name **umabhu**, with no obvious meaning, is a long-established name; whereas **umacutha** is a coinage based on the verb *cutha* 'draw up the body into a still, tense position'.

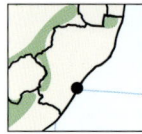

umabhocashile Eurasian Bittern *Botaurus stellaris*
The generic name **umabhu** has been extended with *ochashile* 'which is hidden' in reference to the highly secretive nature of this bird.

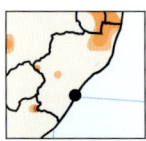

umabhumfashane Dwarf Bittern *Ixobrychus sturmii*
The generic name **umabhu** has been extended with *mfushane* 'short', reflecting the English name.

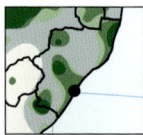

umacuthomncane Little Bittern *Ixobrychus minutus*
The generic name **umacutha** has been extended with *omncane* 'which is little'.

uthekwane Hamerkop *Scopus umbretta*
The name **uthekwane** has been said to be derived from *itheku* 'bay' 'lagoon' with the biological marker *–ane*. The same word – *itheku* – is the base of eThekwini, the Zulu name for Durban.

One of the better-known Zulu bird names, even among non-Zulu-speaking whites of colonial Natal. The bird is praised in Zulu as uThekwane kaZiluba 'Hamerkop, son of Mr Plumes' – a reference to its crest.

This is a bird of great omen in traditional Zulu culture. Flying over the hut calling, and worse sitting on the hut, are signs of great misfortune and evil. The huge nest, constructed with all kinds of cast off human debris, suggests links to *abathakathi* 'witches'. When the bird sits motionless at water's edge, it is believed to be looking at its reflection and bemoaning its ugliness.

The generic name *umacutha* is used for four smaller heron species. It is derived from the verb *cutha* 'draw the body tense'.

umacuthobomvu White-backed Night Heron *Gorsachius leuconotus*
The generic name **umacutha** has been extended with *obomvu* 'which is red'.

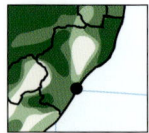

usiba Black-crowned Night Heron *Nycticorax nycticorax*
The word **usiba** means both 'long feather or plume of bird' as well as 'night heron' in Zulu.

umabhocashile Eurasian Bittern

umabhumfashane Dwarf Bittern

umacuthomncane Little Bittern ♂

uthekwane Hamerkop

umacuthobomvu White-backed Night Heron

usiba Black-crowned Night Heron

HERON AND EGRET

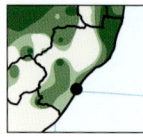
umacuthomhlophe Squacco Heron *Ardeola ralloides*
The generic name **umacutha** has been extended with *omhlophe* 'which is white'.

An alternative name **intolibantshi** has been recorded orally (not in the workshops) but not confirmed. This word for a waistcoat is adopted from Afrikaans *onderbaadjie*. It could be said that the bird appears to be wearing a waistcoat.

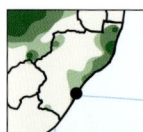
unogilonkomnyama Black Heron *Egretta ardesiaca*
The generic name **unogilonki** (see below) has been extended with *omnyama* 'which is black'.

EGRET GENERIC NAME: ilanda, ingekle . Two generic names are used interchangeably for the cluster of white herons and/or egrets in Zulu folk taxonomy.

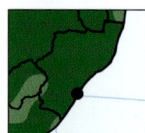
ilanda, umlindankomo Western Cattle Egret *Bubulcus ibis*
Both **ilanda** and **umlindankomo** are equally well-known names for the Cattle Egret (colloquially known in English as 'tick-bird'). **ilanda** is derived from the verb *landa* 'follow' and **umlindankomo** means 'what attends the cattle'.

ingeklenkulu Great Egret *Ardea alba*
The generic name **ingekle** is extended with *enkulu* 'which is big'.

umanyatheludaka Intermediate Egret *Egretta intermedia*
This name is derived from –*ma*– 'characteristically', with the verb *nyathela* 'tread', 'tread on', and *udaka* 'mud' and is a reference to the dark colouring of the lower portion of the leg, seen as a distinctive feature of this species of egret.

ingeklencane Little Egret *Egretta garzetta*
The generic name **ingekle** has been extended with *encane* 'which is small'.

ilanda Western Cattle Egret feeding on insects disturbed by cattle

umacuthomhlophe Squacco Heron (non-br)

unogilonkomnyama Black Heron

ilanda Western Cattle Egret

ingeklenkulu Great Egret

umanyatheludaka Intermediate Egret

ingeklencane Little Egret

HERON

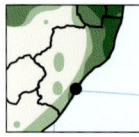
umacutholuhlaza Striated Heron *Butorides striata*

The generic name **umacutha** has been extended with *oluhlaza* 'which is green', reflecting the earlier English name Green-backed Heron.

HERON GENERIC NAME: unokilonki .After considerable discussion, consensus was reached on using **unokilonki** as a generic name for herons, and this generic has been extended for some of the species below.

unokilonkojwayelekile Grey Heron *Ardea cinerea*

The generic name **unokilonki** has been extended with *ojwayelekile* 'which is common', 'which is well-known'.

unokilonkolikhandamnyama Black-headed Heron *Ardea melanocephala*

The generic name **unokilonki** has been extended with *elikhandamnyama* 'black-headed', from *ikhanda* 'head' and *mnyama* 'black'.

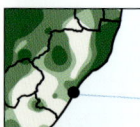
unozalizingwenya Goliath Heron *Ardea goliath*

This name means 'that which gives birth to crocodiles', from *zala* 'give birth' and *izingwenya* 'crocodiles', and refers to this bird feeding in the shallow waters of pans in the company of young basking crocodiles.

unokhoboyi Purple Heron *Ardea pupurea*

The name **unokhoboyi** has been used for the Goliath Heron but is assigned here to the Purple Heron. There is no obvious underlying meaning.

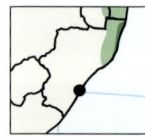
umacuthomnyama Rufous-bellied Heron *Butorides rufiventris*

This name is derived from **umacutha** with the extension *omnyama* 'which is black'. Apart from its rufous belly, the bird is distinctly black.

unokilonkojwayelekile Grey Heron feeding

umacutholuhlaza Striated Heron

unokilonkojwayelekile Grey Heron

unokilonkolikhandamnyama Black-headed Heron

unozalizingwenya Goliath Heron

unokhoboyi Purple Heron

umacuthomnyama Rufous-bellied Heron

GANNET, CORMORANT AND DARTER

isicibamanzi Cape Gannet *Morus capensis*
This name is based on *ciba* 'move like an arrow' and *amanzi* 'water' and provides an excellent picture of a gannet plunging into the water from on high.

CORMORANT GENERIC NAME: iwonde The name **iwonde**, with no obvious underlying meaning, is generally well-known for the cormorant cluster and is used as a generic name here.

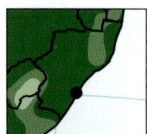

iphishamanzi Reed Cormorant *Microcarbo africanus*
The name **iphishamanzi**, meaning 'what breaks water', is still in common usage for this species.

iwondelimhlophe White-breasted Cormorant *Phalacrocorax lucidus*
The generic name **iwonde** has been extended with *elimhlophe* 'which is white'.

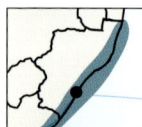

iwonde lasolwandle Cape Cormorant *Phalacrocorax capensis*
For this bird, the only marine cormorant in KwaZulu-Natal, the generic name **iwonde** has been extended with *lasolwandle* 'of the sea'.

inyoninyoka African Darter *Anhinga rufa*
The name **inyoninyoka** is a compound of *inyoni* 'bird' and *inyoka* 'snake' and refers to the bird's snakelike neck. This bird is often informally known in English as the 'Snake Bird' for this reason.

inyoninyoka African Darter eating large fish

isicibamanzi Cape Gannet

iphishamanzi Reed Cormorant

iphishamanzi Reed Cormorant (non-br)

iwondelimhlophe White-breasted Cormorant

iwonde lasolwandle Cape Cormorant

inyoninyoka African Darter

OSPREY, EAGLE, VULTURE, CUCKOO-HAWK, BUZZARD AND HAWK

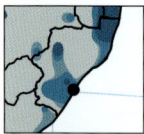

inkwazana Western Osprey *Pandion haliaetus*
The diminutive suffix *–ana* has been added to the name **inkwazi** 'fish eagle' to form the name of this species.

inkwazi African Fish Eagle *Haliaeetus vocifer*
The widely-known and long-established name **inkwazi** is simply the generic name **ukhozi** used for all eagles, in a different noun class, and with the vowel /**a**/ inserted between /**kho**/ and /**zi**/: thus **in** plus **k[h]o** plus **a** plus **zi** = **inkwazi**.

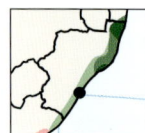

inqemvuma Palm-nut Vulture *Gypohierax angolensis*
This name is formed simply by juxtaposing the nouns **inqe** 'vulture' and *(u)mvuma* 'Raphia Palm'.

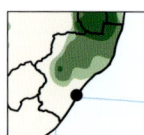

usomthende African Cuckoo-Hawk *Aviceda cuculoides*
This name is based on the noun *umthende* 'stripe', prefixed with *–so–*. Loosely translated, the name means 'the well-striped one'.

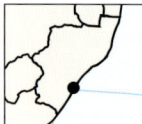

umanyovini European Honey Buzzard *Pernis apivorus*
The name *umnyovu* is composed of *–ma–* 'characteristically' with *[e]nyovini* 'amongst the wasps', from *unyovu* 'wasp'. The name is thus a reference to the bird's diet.

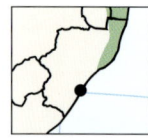

umahlwithilulwane Bat Hawk *Macheiramphus alcinu*
This name is a compound of *–ma–* 'characteristically', with *hlwitha* 'snatch away' and *ilulwane* 'bat'.

inqemvuma Palm-nut Vulture eating raphia palm fruit

inqemvuma Palm-nut Vulture nest high in raphia palm

inkwazana Western Osprey

inkwazi African Fish Eagle

inqemvuma Palm-nut Vulture

usomthende African Cuckoo-Hawk

umanyovini European Honey Buzzard (juv)

umahlwithilulwane Bat Hawk

HARRIER-HAWK, KITE AND SECRETARYBIRD

ijikanyawo African Harrier-Hawk *Polyboroides typus*
This name is a compound of *jika* 'turn' and *unyawo* 'foot', referring to the way the feet of this bird can bend forwards, backwards and sideways, allowing the bird to reach into holes. This characteristic feature of the bird is seen as being more salient than the bare cheeks which gave the bird its previous name Gymnogene.

udemezane, umazabelweni Black-winged Kite *Elanus caeruleus*
The name **udemezane**, with no obvious underlying meaning, is a well-known name for this bird. The lesser-known name **umazabelweni** is derived from the noun *izabelo*, meaning a wide open area where cattle graze, exactly the sort of environment this bird prefers.

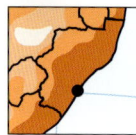

unhloyile Yellow-billed Kite *Milvus aegyptius*
The Yellow-billed Kite has many Zulu names. The bird is greatly loved, and is praised as Nhloyile kaGelegele 'Kite, son of Whirlwind'.

The name uNhloyile is used as an alternative to the name uNcwaba, the Zulu lunar month that runs from mid-July to mid-August. The kite is also the base of the name *idumbe likanhloyile* 'yellow-billed kite's tuber' – *Scadoxus puniceus* or the Snake Lily, which emerges from the ground at the same time the kite returns from wintering in the north.

The kite plays the role of a 'tooth fairy' in traditional Zulu culture. Children who have lost a tooth throw down the old one in front of a kite and chant "*Nhloyile, Nhloyile! Thatha lelizinyo lami elidala ungiph' elisha!*" 'Kite, Kite! take this old tooth of mine and give me a new one!'.

Alternative names still in use: **ukholwe**, **ukholo**, and **isikhokhwane**. None of these have an obvious underlying meaning.

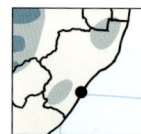

unhloyile waseYurobhu Black Kite *Milvus migrans*
This kite is similar to the Yellow-billed Kite in appearance, so the name of that bird was extended with *waseYurobhu* 'from Europe' for this uncommon migrant species.

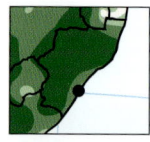

intinginono, intungunono Secretarybird *Sagittarius serpentarius*
Both forms of the Zulu name are equally well-known and in widespread use. There is no obvious underlying meaning.

udemezane Black-winged Kite feeding on mouse

VULTURE

VULTURE GENERIC NAME: inqe The name **inqe** for vultures in general is long-established and well-known. A less-used generic name for vultures is **idlanga**.

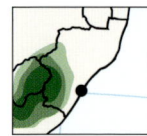

ukhozilwentshebe Bearded Vulture *Gypaetus barbatus*
The generic name **ukhozi**, used for eagles, is extended with *lwentshebe* 'of the beard', thus in Zulu this bird is a 'bearded eagle' rather than a 'bearded vulture'.

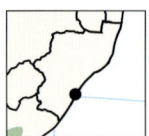

uphalane Egyptian Vulture *Neophron percnopterus*
Still plentiful in KwaZulu-Natal at the time of kings Shaka, Dingane and Mpande in the 19th century, by mid-20th century bird guides were saying "extremely rare" or "no longer found in this area". As these birds always flew in pairs, the name **uphalane** was applied to twin regiments that fought together and shared the same barracks. Another name for this bird is **unobhongoza**, and one of Dingane's regiments was known as the uNobhongoza.

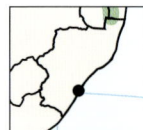

inqelincane Hooded Vulture *Necrosyrtes monachus*
The generic name **inqe** has been extended with *elincane* 'which is small'.

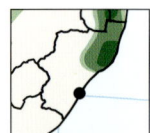

inqe lehlanze White-backed Vulture *Gyps africanus*
The generic name **inqe** has been extended with *lehlanze* 'of the open plain'.

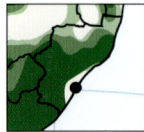

idlanga lentaba Cape Vulture *Gyps coprotheres*
The alternative generic name **idlanga** has been extended with *lentaba* 'of the mountain' to distinguish this bird from **indlangamandla** the Lappet-faced Vulture (see below).

inqe lehlanze White-backed Vultures feeding on Impala

ukhozilwentshebe Bearded Vulture

ukhozilwentshebe Bearded Vulture (juv)

uphalane Egyptian Vulture

inqelincane Hooded Vulture

inqe lehlanze White-backed Vulture

idlanga lentaba Cape Vulture

VULTURE, BATELEUR AND SNAKE EAGLE

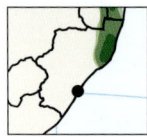

ukhandelimhlophe White-headed Vulture *Trigonoceps occipitalis*
This name is a compound of *ikhanda* 'head' and *elimhlophe* 'which is white'.

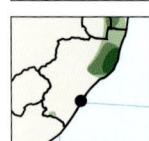

indlangamandla Lappet-faced Vulture *Torgos tracheiotos*
This name has a double meaning: it is simultaneously a compound of *dla* 'eat' and *ngamandla* 'with great strength', and **idlanga** 'vulture' with *amandla* 'strength'. The bird is thus in this name a powerful vulture that feeds voraciously.

EAGLE GENERIC NAME: ukhozi. The generic name **ukhozi** for eagles is a very old Bantu-language word, and **ukhozi** in various forms is used for raptors of various kinds – both large and small – in Bantu languages from eastern southern Africa to Central and East Africa.

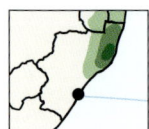

ingqungqulu Bateleur *Terathopius ecaudatus*
This is a well-known name for this bird. The bateleur plays an important role in Zulu praise poetry and in traditional beliefs as a bird of war and battle. The syllables *ngqu ngqu* of its name have been linked both to the clapping of its own wings and to the sound of distant drums or gunfire. The bateleur also has the praise-name **indlamadoda** 'what eats men' which is said to be from its habit of eating the corpses of those slain in battle.

Snake eagle generic name: the name **indlanyoka** simply means 'what eats the snake'. In the three species listed below, this generic term is further distinguished with a colour term.

indlanyokemnyama Black-chested Snake Eagle *Circaetus pectoralis*
The generic name **indlanyoka** is extended with *emnyama* 'which is black'.

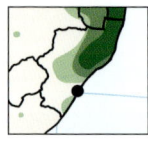

indlanyokensundu Brown Snake Eagle *Circaetus cinereus*
The generic name **indlanyoka** has been extended with *ensundu* 'which is brown'.

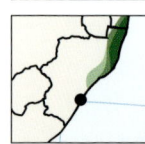

indlanyokempunga Southern Banded Snake Eagle *Circaetus fasciolatus*
The generic name **indlanyoka** has been extended with *empunga* 'which is grey'.

ingqungqulu Bateleur (juv)

indlanyokempunga Southern Banded Snake Eagle

ukhandelimhlophe White-headed Vulture

indlangamandla Lappet-faced Vulture

ingqungqulu Bateleur ♀

indlanyokemnyama Black-chested Snake Eagle

indlanyokensundu Brown Snake Eagle

indlanyokempunga Southern Banded Snake Eagle

HARRIER

SMALLER RAPTORS GENERIC NAMES:

Harriers: umamhlangeni: Previously recorded only for the African Marsh Harrier, this name has been broadened into a generic for all harriers. The name is derived from *–ma–* and the locative form of *umhlanga* 'reed bed' so it means 'that characteristically [found] in a reed bed'.

Goshawks and Hobby: uheshe: This name is derived from the verb *hesha* 'swoop down upon', 'snatch away'.

Falcons and Buzzards: uklebe: This name traditionally refers to any small kind of hawk, given to devouring young fowls.

Sparrowhawks: uheshane This name is derived from **uheshe** with the suffix *–ane*, used for indicating biological species but also suggesting the diminutive.

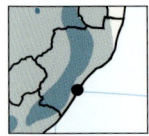

umamhlangeni wasentshonalanga Western Marsh Harrier *Circus aeruginosus*

The generic name **umamhlangeni** has been extended with *wasentshonalanga* 'from the west'.

umamhlangenonsundu African Marsh Harrier *Circus ranivorus*

The generic name **umamhlangeni** has been extended with *onsundu* 'which is brown'.

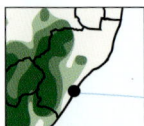

umamhlangenomnyama Black Harrier *Circus maurus*

The generic name **umamhlangeni** has been extended with *omnyama* 'which is black'.

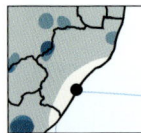

umamhlangenomphunga Pallid Harrier *Circus macrourus*

The generic name **umamhlangeni** has been extended with *omphunga* 'which is grey'.

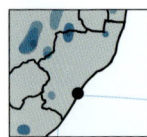

unohlohlweni Montagu's Harrier *Circus pyargus*

The name **unohlohlweni** is in common use, with an unknown meaning.

umamhlangenonsundu African Marsh Harrier (juv)

umamhlangeni wasentshonalanga Western Marsh Harrier (imm)   **umamhlangenonsundu** African Marsh Harrier

umamhlangenomnyama Black Harrier   **umamhlangenomphunga** Pallid Harrier ♂

unohlohlweni Montagu's Harrier ♂   **unohlohlweni** Montagu's Harrier (juv)

GOSHAWK, SHIKRA AND SPARROWHAWK

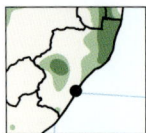

uheshomlotha Gabar Goshawk *Micronisus gabar*
The generic name **uheshe** has been extended with *omlotha* 'ashen', 'grey'.

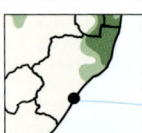

uheshomidwayidwa Shikra *Accipiter badius*
The generic name **uheshe** has been extended with *midwayidwa* 'striped'.

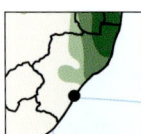

usozi Lizard Buzzard *Kaupifalco monogrammicus*
The name **usozi** refers to its whistling sound, with –so– prefixed to an onomatopoeic 'zi-i-i-i'.

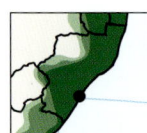

usomheshe, imvumvuyane African Goshawk *Accipiter tachiro*
The name **usomheshe**, where the prefix –so– is added to the generic word for goshawks, and is used in the sense of 'father of', indicates the larger and more powerful of all the goshawks.

The name **imvumvuyane** is regionally well-known. The name is likely derived from the suffix –ane (used for biological species) with the noun *imvumvu* 'powdery, grey wood ash', a reference to the bird's coloration.

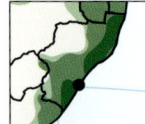

uheshanyana, umqwayini Little Sparrowhawk *Accipiter minullus*
The generic **uheshane**, already a diminutive, has another diminutive suffix –ana attached to form **uheshanyana**. The name **umqwayini** is well-known, and may be linked to the Zulu ideophone *qwáyi* meaning 'in an eye-blink', and possibly referring to the extremely rapid appearance and disappearance of this bird.

uheshanobomvu Rufous-breasted Sparrowhawk *Accipiter rufiventris*
The generic name **uheshane** has been extended with *obomvu* 'which is red'.

uheshomlotha Gabar Goshawk (juv)

uheshomlotha Gabar Goshawk

uheshomidwayidwa Shikra

usozi Lizard Buzzard

usomheshe African Goshawk

uheshanyana Little Sparrowhawk

uheshanobomvu Rufous-breasted Sparrowhawk

SPARROWHAWK, BUZZARD & EAGLE

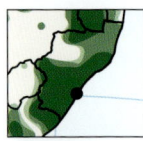

uheshanomnyama Black Sparrowhawk *Accipiter melanoleucus*
The generic name **uheshane** has been extended with *omnyama* 'which is black'.

isanxa Common Buzzard *Buteo buteo*
This name is based on *–sa–* 'something like' with *inxa* 'edge, margin'. The bird is commonly found on the edges of wooded thickets.

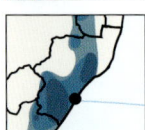

uklebe lwehlathi Forest Buzzard *Buteo trizonatus*
The generic name **uklebe** has been extended with *lwehlathi* 'of the forest' for this species.

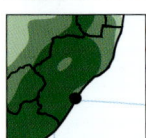

inhlandlokazi Jackal Buzzard *Buteo rufofuscus*
This name is well-established and widely known for this bird. It has no clear underlying meaning.

See (page 62) for the use of *ukhozi* as a generic name for eagle.

ukhozolunsundu Tawny Eagle *Aquila rapax*
The generic name **ukhozi** has been extended with *olunsundu* 'which is brown'.

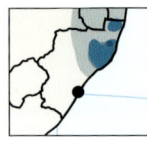

ukhozolumabala Lesser Spotted Eagle *Clanga pomarina*
The generic name **ukhozi** has been extended with *olumabala* 'which has spots'.

isanxa Common Buzzard

inhlandlokazi Jackal Buzzard

uheshanomnyama Black Sparrowhawk

isanxa Common Buzzard

uklebe lwehlathi Forest Buzzard

inhlandlokazi Jackal Buzzard

Adam Riley - Rockjumper Birding

ukhozolunsundu Tawny Eagle

ukhozolumabala Lesser Spotted Eagle (juv)

EAGLE & HAWK-EAGLE

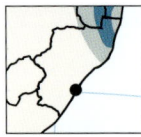

ukhozimuhlwa Steppe Eagle *Aquila nipalensis*
This name, compounded from **ukhozi** and *[u]muhlwa* 'termites', refers to the bird's diet.

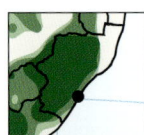

ukhozolumnyama Verreaux's Eagle *Aquila verreauxii*
The generic name **ukhozi** has been extended with *olumnyama* 'which is black'.

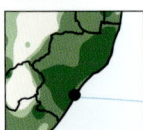

inkosiyezinkozi, isihuhwa Martial Eagle *Polemaetus bellicosus*
The name **inkosiyezinkozi**, meaning 'chief of the eagles', reflects the pre-eminent status of this eagle amongst eagles generally. The name **isihuhwa** is a praise name for both the Martial Eagle and the Crowned Eagle.

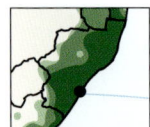

ukhoziqholwane, isihuhwa Crowned Eagle *Stephanoaetus coronatus*
The generic name **ukhozi** is extended with *–qholwane* 'crested', 'crowned'. The praise-name **isihuhwa** has been retained for this bird.

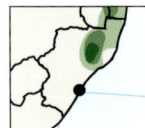

ukhozolumidwayidwa African Hawk-Eagle *Aquila spilogaster*
The generic name **ukhozi** was extended with *olumidwayidwa* 'which has many streaks'.

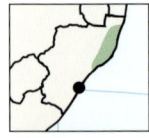

ukhozolumabalabala Ayres's Hawk-Eagle *Hieraaetus ayresii*
The generic name **ukhozi** was extended with *olumabalabala* 'which has many markings'.

ukhoziqholwane Crowned Eagle with monkey prey

ukhozimuhlwa Steppe Eagle

ukhozolumnyama Verreaux's Eagle

inkosiyezinkozi Martial Eagle

ukhoziqholwane Crowned Eagle

ukhozolumidwayidwa African Hawk-Eagle

ukhozolumabalabala Ayres's Hawk-Eagle

EAGLE & KESTREL

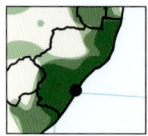
isiphungumangathi Long-crested Eagle *Lophaetus occipitalis*
This is a well-known name for this bird. The Long-crested Eagle is known for finding lost cattle. When a herd boy finds that his cattle have wandered off, he seeks out a Long-crested Eagle and asks of it *"Siphungumangathi, siphungumangathi –zilaphi izinkomo zikababa?"* 'Siphungumangathi, siphungumangathi – where are my father's cattle?' upon which the eagle points its crest in the direction of the strayed cattle. The bird shares both the name and the cattle-finding function with any insect chrysalis with one rounded and one pointed end.

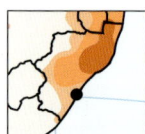
ukhozolusisila Wahlberg's Eagle *Hieraaetus wahlbergi*
The generic name **ukhozi** has been extended with *olusisila* < *isisila* 'bird's tail' in reference to the longer tail of this eagle.

ukhozolumadladla Booted Eagle *Hieraaetus pennatus*
The generic name **ukhozi** has been extended with *olumadladla* 'which has shaggy growth of hair'. Interestingly, the Southern Sotho name for the Booted Eagle – *Ntsu-marikhoana* – is derived from *marikhoe*, the plural of *borikhoe* 'trousers', borrowed from Afrikaans *broek*.

KESTREL GENERIC NAME: umathebethebana The name **umathebethebana** is well-known. It has no obvious underlying meaning.

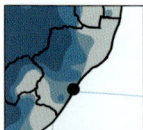
umathebethebanomncane Lesser Kestrel *Falco naumanni*
The generic **umathebethebana** has been extended with *omncane* 'which is small[er]'.

umathebethebana wamadwala Rock Kestrel *Falco rupicolus*
The generic **umathebethebana** has been extended with *wamadwala* 'of the rocks'. The previously recorded names **utebetebana** and **umathebeni** are regional variations of the generic name, and are among several variations which have been recorded and are still used orally.

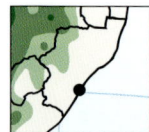
umathebethebanomkhulu Greater Kestrel *Falco rupicoloides*
The generic **umathebethebana** has been extended with *omkhulu* 'which is large[r]'.

isiphungumangathi Long-crested Eagle in flight

isiphungumangathi Long-crested Eagle

ukhozolusisila Wahlberg's Eagle

ukhozolumadladla Booted Eagle

umathebethebanomncane Lesser Kestrel

umathebethebana wamadwala Rock Kestrel

umathebethebanomkhulu Greater Kestrel

FALCON & HOBBY

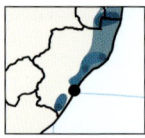

uklebompunga Sooty Falcon *Falco concolor*
The generic name **uklebe** has been extended with *punga* 'which is grey'.

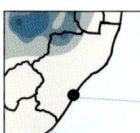

uklebonyawobomvu Red-footed Falcon *Falco vespertinus*
The generic name **uklebe** has been extended with *onyawobomvu* 'which is red-footed'.

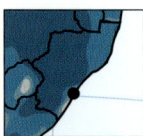

oklebeklebe Amur Falcon *Falco amurensis*
The generic name **uklebe** has been reduplicated, and the name, unusually, put into the plural form, to indicate congregating in vast numbers.

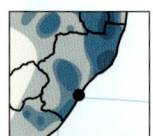

uklebosankonjane Eurasian Hobby *Falco subbuteo*
The generic name **uklebe** has been extended with *osankonjane*, derived from *–sa–* 'something like' and *inkonjane* 'swallow'.

uklebemawa Lanner Falcon *Falco biarmicus*
The generic name **uklebe** has been extended with *[a]mawa* 'cliffs' in reference to its preferred nesting and roosting place.

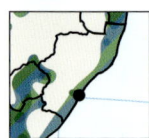

uklebosikhweshekweshe Peregrine Falcon *Falco peregrinus*
The generic name **uklebe** has been extended with *osikhweshekhweshe*, a reduplication of the ideophone *kwéshe* 'of light, quick movement'.

oklebeklebe Amur Falcon ♀

uklebompunga Sooty Falcon

uklebonyawobomvu Red-footed Falcon

oklebeklebe Amur Falcon ♂

uklebosankonjane Eurasian Hobby

uklebemawa Lanner Falcon

uklebosikhweshekweshe Peregrine Falcon

BUSTARD & KORHAAN

BUSTARD AND KORHAAN GENERIC NAMES: The **bustards** and the **korhaans** are regarded as distinctive enough in the Zulu folk taxonomy to have long had distinct species-specific names so there is no generic term, and there has been no need to extend a generic name to make species distinctions.

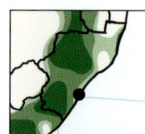

iseme Denham's Bustard *Neotis denhami*
This name is well-known and in widespread use, and has no obvious underlying meaning.

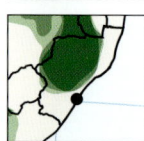

inkakulo White-bellied Bustard *Eupodotis senegalensis*
The name **inkakulo** is in current use today for this species, and has no obvious underlying meaning.

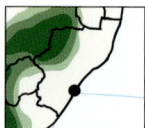

umbhukwane Blue Korhaan *Eupodotis caerulescens*
A well-known name for this bird, with no obvious underlying meaning.

umngqithi Red-crested Korhaan *Eupodotis ruficrista*
The name **umngqithi** is in current use for the Red-crested Korhaan, and its underlying meaning is not known.

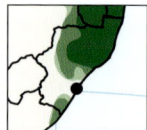

ufumba Black-bellied Bustard *Lissotis melanogaster*
The name **ufumba** is well-known for this species and has no known underlying meaning.

An older, unrecorded name for this bird, still in use in some parts of KwaZulu-Natal is the name **umshukwane**, derived from the verb *shuka* 'make a motion of rubbing clothes together when washing them', a reference to the bird's display of raising and lowering its head.

umngqithi Red-crested Korhaan ♀

umngqithi Red-crested Korhaan with crest raised ♂

iseme Denham's Bustard ♂

inkakulo White-bellied Bustard ♂

umbhukwane Blue Korhaan ♂

umngqithi Red-crested Korhaan ♂

ufumba Black-bellied Bustard ♂

ufumba Black-bellied Bustard ♀

FLUFFTAIL & CRAKE

FLUFFTAIL GENERIC NAME: ubhavuzile It is possible that this name is derived from the ideophone *bhávu* 'of making a noise on a paraffin-tin'. A verb derived from this ideophone would be *bhavuza*, and in the perfect tense this would be *bhavuzile*, with the approximate meaning 'having made a noise sounding like the resonance of an empty paraffin tin'.

ubhavuzilobomvana Red-chested Flufftail *Sarothrura rufa*

The generic name **ubhavuzile** has been extended with *obomvana* 'which is red'. The diminutive suffix *–ana* has been added, giving 'little red flufftail'.

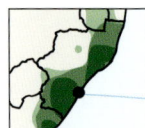

ubhavuzile wehlathi Buff-spotted Flufftail *Sarothrura elegans*

The generic name **ubhavuzile** has been extended with *wehlathi* 'of the forest'.

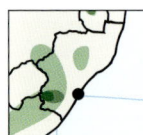

ubhavuzilomidwayidwa Striped Flufftail *Sarothrura affinis*

The generic name **ubhavuzile** has been extended with *omidwayidwa* 'which is striped'.

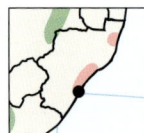

ubhavuzilomhlophe White-winged Flufftail *Sarothrura ayresii*

The generic name **ubhavuzile** has been extended with *omhlophe* 'which is white'.

umjekejeke waseAfrika African Crake *Crex egregia*

The name **umjekejeke** is well-known for crakes, and has the regional variation **umjengejenge**. Umjekejeke is extended here with *waseAfrika* 'from Africa' to distinguish it from the following species.

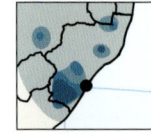

umjekejeke wasenyakatho Corn Crake *Crex crex*

The name **umjekejeke** has been extended with *wasenyakatho* 'from the north' i.e. Europe, to distinguish it from the previous species.

ubhavuzile wehlathi Buff-spotted Flufftail ♀ turning eggs. Nest at ground level.

ubhavuzilobomvana Red-chested Flufftail ♂

ubhavuzile wehlathi Buff-spotted Flufftail ♂

ubhavuzilomidwayidwa Striped Flufftail ♂

ubhavuzilomhlophe White-winged Flufftail ♂

umjekejeke waseAfrika African Crake

umjekejeke wasenyakatho Corn Crake

CRAKE, RAIL, MOORHEN & FINFOOT

isizinzana Baillon's Crake *Porzana pusilla*
The name **isizinzi** is used generally for some crakes and rails. Here it is extended with the diminutive *–ana*.

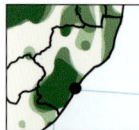

isizinzi African Rail *Rallus caerulescens*
This is a well-known name for this bird, possibly linked to the ideophone *zínzi* 'of settling down', referring to the way in which this bird settles down in a dense reed-bed.

inkukhuyamanzi Common Moorhen *Gallinula chloropus*
This coinage is a compound of *inkukhu* 'chicken', 'fowl' and *yamanzi* 'of the water'. It is interesting to note the similarity behind names in other southern Bantu languages: the North Sotho name *kgogomeetse* 'chicken [of the] water', the Northern and Southern Sotho *kgogonoka* and *khohonoka* 'chicken [of the] river', and the Tsonga *kukumezane* 'little chicken bird'.

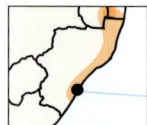

inkukhumezane Lesser Moorhen *Paragallinula angulata*
This bird name is well-known orally. See the Tsonga name for the Moorhen in the entry above.

igwedlamanzi African Finfoot *Podica senegalensis*
This name is a compound of *gwedla* 'paddle' and *amanzi* 'water'. The underlying meaning is identical to the Afrikaans name *watertrapper* 'water-treader'.

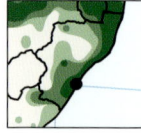

isiqhanazana Black Crake *Amaureornis flavirostris*
This name is derived from the verb *qananaza*, with the meaning of 'run along quickly, trip along'. A possible alternative name is **umanyenyane**, described as 'a small bird seen running on the water-lily leaves'.

igwedlamanzi African Finfoot ♀

isizinzana Baillon's Crake

isizinzi African Rail

inkukhuyamanzi Common Moorhen

inkukhumezane Lesser Moorhen ♀

igwedlamanzi African Finfoot ♂

isiqhanazana Black Crake

GALLINULE, SWAMPHEN, COOT & CRANE

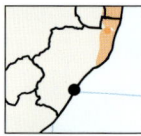

unomhlangomncane Allen's Gallinule *Porphyria alleni*
This bird, like that of the African Swamphen (see below) is based on the noun *umhlanga* 'reed-bed'. Here, *umhlanga* is prefixed with –*no*– and extended with *omncane* 'which is small'. The name thus means 'the small reed-bed bird'.

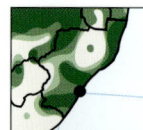

inkukhuyomhlanga African Swamphen *Porphyrio madagascariensis*
This name is a compound of *inkukhu* 'chicken' and *yomhlanga* 'of the reed-bed'.

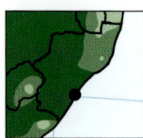

intuntwane Red-knobbed Coot *Fulica cristata*
The previously unrecorded name **intuntwane** is widely-known for this bird.

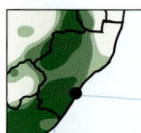

unohemu Grey Crowned Crane *Balearica reguloram*
Various South African languages have onomatopoeia-based names for this bird: Afrikaans *mahem*, Southern Sotho *lehemuhemu*, Xhosa *ihem*, and Zulu **unohemu**.

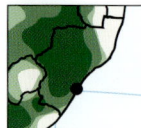

indwe Blue Crane *Grus paradisea*
Various sources have **indwe** for this bird; others have **indwa**. The variant **indwe** is the better known orally in KwaZulu-Natal. North of the uThukela River, however, the more common name is **uzuka**. This name has an interesting history. An earlier meaning is 'sixpenny piece'. When South Africa 'went metric' in 1961, the old sixpence became the new South African 5c piece, and displayed on one side was the national bird of South Africa, i.e. the Blue Crane. The old name of the sixpence transferred to the 5c piece and the name of the coin in turn was passed on to the bird. It is curious that this only happened in the northern half of KwaZulu-Natal, where many Zulu-speaking people see the name **indwe** as being used by 'those in the south'.

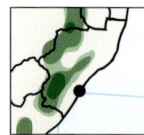

ubhamukwe Wattled Crane *Grus carunculata*
The name **ubhamukwe** is well-known for this bird. It has no obvious underlying meaning.

unohemu Grey Crowned Crane

unomhlangomncane Allen's Gallinule

inkukhuyomhlanga African Swamphen

intuntwane Red-knobbed Coot

unohemu Grey Crowned Crane

indwe Blue Crane

ubhamukwe Wattled Crane

THICK-KNEE, STILT, AVOCET, OYSTERCATCHER & CRAB PLOVER

isiwelewele Water Thick-knee *Burhinus vermiculatus*
The name **isiwelewele** was assigned to this bird, with the underlying meaning 'species of marsh bird, which utters a loud plaintive call'.

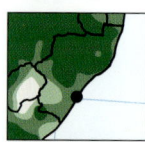

umbangaqhwa Spotted Thick-knee *Burhinus capensis*
The name **umbangaqhwa** is well-known for this bird. It is derived from the Zulu verb *banga* 'cause' and the noun *iqhwa* 'snow', 'frost'. When the bird lies low, with the white spots on its back it looks like a frost-covered rock.

uduku Black-winged Stilt *Himantopus himantopus*
This name, in common use for the Black-winged Stilt, is said to be onomatopoeic.

usipheshula Pied Avocet *Recurvirostra avosetta*
This name is derived from the noun *isipheshula* 'anything turned up' in reference to the distinctive bill of this bird.

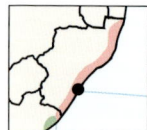

unozila African Oystercatcher *Haematopus moquini*
The all-black African Oystercatcher is perceived as a bird in mourning, hence the name **unozila**, derived from –no– and the verb *zila* 'to mourn'.

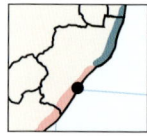

umandlankala Crab Plover *Dromas ardeola*
The name **umandlankala** is based on –ma– 'characteristically' with the verb *dla* 'eat' and the abbreviated form *[i]nkala of inkalankala* 'crab'. There is a village named eMandlankala between eSikhawini and Richards Bay on the KwaZulu-Natal North Coast where these birds occur.

uduku Black-winged Stilt (juv)

isiwelewele Water Thick-knee

umbangaqhwa Spotted Thick-knee

uduku Black-winged Stilt

usipheshula Pied Avocet

unozila African Oystercatcher

umandlankala Crab Plover (juv)

LAPWING & PLOVER

PLOVER/LAPWING NAME: There is no single generic name for the plover/lapwing group. It is of interest to note that the name **isibulalambiza** 'what breaks the clay pot' means a 'species of small bird which draws people from its nest through feigning inability to fly'. The name **isibulalambiza** may well have been a historical name for this group of birds, as this 'broken-wing display' is typical of plovers and lapwings.

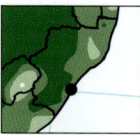

indudumela Blacksmith Lapwing *Vanellus armatus*

The name **indudumela** is well-known for this bird. The name is likely linked to the *klink klink klink* hammer-on-anvil call of this bird.

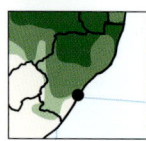

umadevaphuzi African Wattled Lapwing *Vanellus senegallus*

The name **umadavuphuzi** is a noun + adjective construction which means 'having a yellow moustache'.

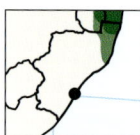

umahambehlala Senegal Lapwing *Vanellus lugubris*

The name **umahambehlala** consists of the prefix –*ma*– 'characteristically' with the verbs *hamba* 'go' and *hlala* 'stay', giving the name the underlying meaning of 'the one that characteristically goes and stays', in reference to the nomadic behaviour of this species.

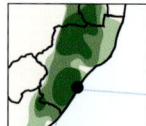

ititihoye Black-winged Lapwing *Vanellus melanopterus*

Out of all the plovers and lapwings that are loosely known as '*ititihoye*', this species is the 'real titihoye', the one that calls out its own name. The name **ititihoye** was made famous by Alan Paton in the opening paragraphs of his classic book *Cry, The Beloved Country*.

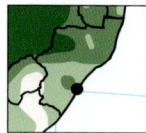

ititihoyenomqhele Crowned Lapwing *Vanellus coronatus*

The name **ititihoye** has been extended with *enomqhele* 'which has a crown'.

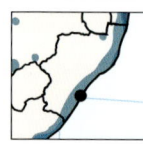

umakhwaphamnyama Grey Plover *Pluvialis squatarola*

The name **umakhwaphamnyama** literally means 'black armpits'.

ititihoye Black-winged Lapwing in flight

indudumela Blacksmith Lapwing

umadevaphuzi African Wattled Lapwing

umahambehlala Senegal Lapwing

ititihoye Black-winged Lapwing

ititihoyenomqhele Crowned Lapwing

umakhwaphamnyama Grey Plover (non-br)

PLOVER

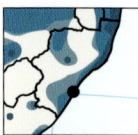
unosongo Common Ringed Plover *Charadrius hiaticula*
This name is based on the prefix *–no–* and the noun *isongo* 'metal armlet'.

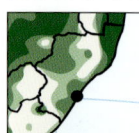
umatatazela Kittlitz's Plover *Charadrius pecuarius*
The name **umatatazela** is based on *–ma–* 'characteristically' with the verb *tatazela* 'be agitated', 'act in a flurried manner', and refers to the agitated movements of the bird as it forages on shorelines of lakes and dams.

igwigwi Three-banded Plover *Charadrius tricollaris*
The name **igwigwi** suggests onomatopoeia as an origin.

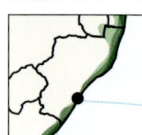
umathantatha White-fronted Plover *Charadrius marginatus*
This name is based on *–ma–* 'characteristically' and the verb *thantatha* 'deviate', 'walk off the path'. A typical behaviour of the bird to run ahead and then deviate to the side on a beach.

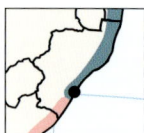
umagevuzomncane Lesser Sand Plover *Charadrius mongolus*
This name is based on *–ma–* 'characteristically' and the verb *gevuza* 'chatter, talk incessantly'. The name is extended with *omncane* 'small' to distinguish this species from the following species.

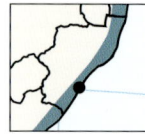
umagevuzomkhulu Greater Sand Plover *Charadrius leschenaultii*
This name is based on *–ma–* 'characteristically' and the verb *gevuza* 'chatter, talk incessantly'. The name is extended with *omkhulu* 'big' to distinguish this species from the previous species.

unosongo Common Ringed Plover eating polychaete worm.

unosongo Common Ringed Plover

umatatazela Kittlitz's Plover

igwigwi Three-banded Plover

umathantatha White-fronted Plover

umagevuzomncane Lesser Sand Plover

umagevuzomkhulu Greater Sand Plover

SNIPE, JACANA, GODWIT & WHIMBREL

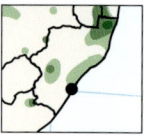

umakhwaneni Greater Painted Snipe *Rostratula benghalensis*
This name could be interpreted in two ways: either as (1) *u + ma + ekhwaneni* from *ikhwane* 'sedge, rush, *Cyperus* sp.', or as (2) *u + (e)makhwaneni* 'among the rushes'. Either way it means the same thing: this is a bird characteristically found among sedge and rushes.

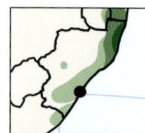

unondwayizomncane Lesser Jacana *Microparra capensis*
The well-known name for the Jacana – **unondwayiza** – is extended with *omncane* 'small'.

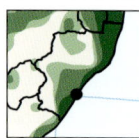

unondwayizomkhulu, umathandaluziba African Jacana *Actophilornis africanus*
The well-known name for the Jacana – **unondwayiza** – is extended with *omkhulu* 'big' to distinguish it from the preceding species. The name **unondwayiza** is based on the verb *dwayiza* 'walk with long strides', and the name uNondwayiza is often used as a nickname for long-legged people.

The name **umathandaluziba** 'the one that likes the lily pads' has also been assigned to this bird.

unununde African Snipe *Gallinaga nigripennis*
The name **unununde** is well-known for this bird and has no apparent underlying meaning.

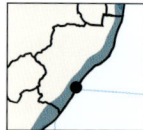

unodaka Bar-tailed Godwit *Limosa lapponica*
This name, derived from *–no–* and *udaka* 'mud', refers to the bird's habits of foraging on exposed mud.

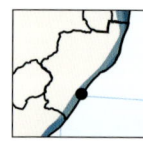

unokhifi Whimbrel *Numenius phaeopus*
The ideophone *khifi* refers to drizzling and misty rain, and is prefixed with *–no–* here. The fine spots and speckles on this bird suggest a light spattering of rain.

unondwayizomkhulu African Jacana (juv)

umakhwaneni Greater Painted Snipe

unondwayizomncane Lesser Jacana

unondwayizomkhulu African Jacana

unununde African Snipe

unodaka Bar-tailed Godwit

unokhifi Whimbrel

REDSHANK, GREENSHANK & SANDPIPER

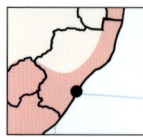

unomlenzobomvu Common Redshank *Tringa totanus*
This name reflects the English vernacular name 'redshank', and is a compound of *umlenze* 'leg' and *obomvu* 'which is red'.

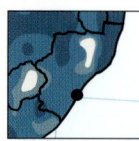

unompempe Common Greenshank *Tringa nebularia*
This name is based on the noun *impempe* 'whistle', prefixed with –*no*–. Impempe Pan in the uMkhuze floodplain is a major destination for birders because of all the different wetland species there.

GENERIC NAME FOR WADERS: osoxhaphozi. The plural word **osoxhaphozi** (sing. **usoxhaphozi**) is derived from the word *ixhaphozi* which refers to wetlands generally, both inland and coastal.

unothwayiza Marsh Sandpiper *Tringa stagnatilis*
This name is formed from the prefix –*no*– and the verb *thwayiza* 'walk with long, swinging steps'.

umakhwifikwifi Wood Sandpiper *Tringa glareola*
This name is derived directly from the adjective *makhwifikhwifi* 'spotted', 'speckled'.

ucijomhlophe Common Sandpiper *Actitis hypoleucus*
This name is based on the noun *umcijo*, a reference to a diamond card in a pack of playing cards, extended with *omhlophe* 'which is white'. The name refers to the narrow white pointed marking on its side.

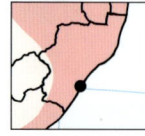

unogqabokhwifi Pectoral Sandpiper *Calidris melanotos*
This name is based on the prefix –*no*–, the ideophone *gqába* 'of marking the face with dots' and *khwifi* 'speckled'.

unompempe Common Greenshank with frog prey

unomlenzobomvu Common Redshank

unompempe Common Greenshank

unothwayiza Marsh Sandpiper

umakhwifikwifi Wood Sandpiper

ucijomhlophe Common Sandpiper

unogqabokhwifi Pectoral Sandpiper

SANDPIPER, SANDERLING, KNOT & STINT

unopheshwana Terek Sandpiper *Xenus cinereus*
The root of this name is the ideophone *phéshu* 'of being turned up'. The ideophone has been prefixed with *–no–* and suffixed with the diminutive *–ana*.

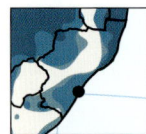

ungozwana Curlew Sandpiper *Calidris ferruginea*
This name is metaphorical in nature, inviting comparison with the tiny elephant shrew (*ingozo*) with its elongated and upturned snout. The diminutive suffix *–ana* has been added.

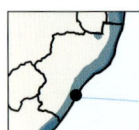

umaphithizela Sanderling *Calidris alba*
This name is based on *–ma–* 'characteristically' with the verb *phithizela* 'move about in a haphazard, confused and disorderly manner'. The reference is to the bird's movements on the wave line.

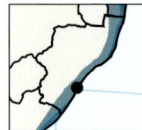

umaphendulamatshe Ruddy Turnstone *Arenaria interpes*
This name is the exact equivalent of the English 'turnstone': it is derived from the verb *phendula* 'turn something over' and *amatshe* 'stones'.

unovimba Red Knot *Calidris canutus*
This is one of the more intriguing coinage, as it relates to the story of King Canute and his attempts to stop or block the tide. The Zulu name reflects this story in the name **unovimba**, from *–no–* and *vimba* 'block off, stem, stop'.

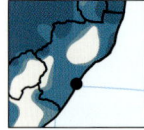

unothwayizana Little Stint *Calidris minuta*
This is the diminutive form of the name **unothwayiza** coined for the Marsh Sandpiper (see above).

ungozwana Curlew Sandpiper extracting a polychaete worm from below the mud surface

unopheshwana Terek Sandpiper

ungozwana Curlew Sandpiper

umaphithizela Sanderling

umaphendulamatshe Ruddy Turnstone

unovimba Red Knot

unothwayizana Little Stint

RUFF, COURSER, PRATINCOLE & GULL

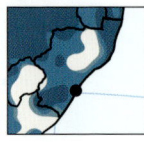

unogqabakazi Ruff ♂ Reeve ♀ *Philomachus pugnax*
This name is based on –no– with *gqaba* 'mark the face, as with incisions' and –*kazi* 'greater'. Although not so common in modern times, such scarification was an important way of recognising one clan from another.

COURSER GENERIC NAME: unobulongwe The name **unobulongwe** is well-known and widely-used for coursers. The name may be derived from *ubulongwe* 'fresh animal dung', with coursers occurring in heavily-overgrazed dung-littered habitats.

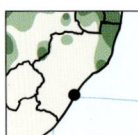

unobulongwana Temminck's Courser *Cursorius temminckii*
As this is the smaller of the two KwaZulu-Natal coursers the generic name **unobulongwe** is suffixed with the diminutive suffix –*ana*.

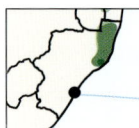

unobulongwonsundu Bronze-winged Courser *Rhinoptilus chalcopterus*
The generic name **unobulongwe** is extended with *onsundu* 'which is brown'.

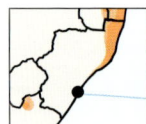

iwamba Collared Pratincole *Glareola pratincola*
The name **iwamba** is well-known and widely-used for this bird.

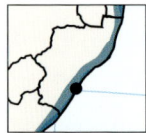

unochweba Kelp Gull *Larus dominicanus*
The name **unochweba** derives from the prefix –*no*– and *ichweba* 'lagoon, bay, expanse of sea'.

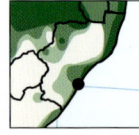

indewula Grey-headed Gull *Chroicocephalus cirrocephalus*
This name for this bird is well-known and widely used. It has been suggested that the name **indewula** refers to the bird 'snatching at flesh like a hyena'.

unochweba Kelp Gull (juv) catching a fish

unogqabakazi Ruff

unobulongwana Temminck's Courser

unobulongwonsundu Bronze-winged Courser

iwamba Collared Pratincole

unochweba Kelp Gull

indewula Grey-headed Gull

TERN

ubhaklakliyo Caspian Tern *Hydroprogne caspia*
This name is based on onomatopoeia. There are two elements here, each containing the rasping guttural sound '*kl*'. The first element (*ikliyo*) has been used in the coining of a number of tern names; to this has been prefixed *–bhakla–*.

ukliyo Lesser-crested Tern *Thalasseus bengalensis*
The name **ukliyo** is onomatopoeic and is used as a base 'root' of the name of several other species of terns.

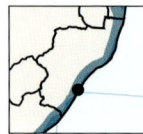
unolwandle Swift Tern *Thalasseus bergii*
This name uses the prefix *–no–* attached to the noun *ulwandle* 'sea'.

unonkliyo Sandwich Tern *Thalasseus sandvicensis*
This is one of a number of names for terns based on the onomatopoeic *kliyo*, with *–no–* prefixed.

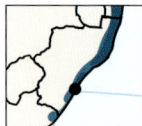
unonklilwane Little Tern *Sternula albifrons*
Another variation on *kliyo*, this time with *–no–* prefixed and *–ane* suffixed.

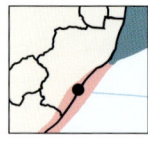
unonkliyomnyama Sooty Tern *Onychoprion fuscatus*
The onomatopoeic element *kliyo* has *–no–* prefixed, and is extended with *omnyama* 'which is black'.

ubhaklakliyo Caspian Tern

ubhaklakliyo Caspian Tern

ukliyo Lesser-crested Tern

unolwandle Swift Tern

unonkliyo Sandwich Tern

unonklilwane Little Tern

unonkliyomnyama Sooty Tern

TERN & SKUA

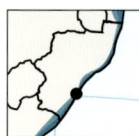
unonteteza Common Tern *Sterna hirundo*
This name is based on the verb *ntenteleza* 'skim along the surface'.

insukakude Arctic Tern *Sterna paradisaea*
This name is based on the verb *suka* 'come from' and the adverb *kude* 'far away'.

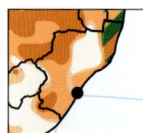
umachibini Whiskered Tern *Chlidonias hybrida*
This name is derived from *emachibini*, the locative form of *amachibi* 'pans, lakes', and refers to the preferred habitat of this bird.

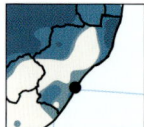
unochibi White-winged Tern *Chlidonias leucopterus*
This name is formed from the prefix *–no–* and the noun *ichibi* 'pan', 'bay', 'lake', 'lagoon'.

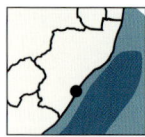
impisiyolwandle Brown Skua *Stercorarius antarcticus*
This name is derived from *impisi* 'hyena' and *yolwandle* 'of the sea', and is a reference to the way this bird steals food items from other birds.

umachibini Whiskered Tern (non-br) on nest with eggs

umachibini Whiskered Tern (partial br) at nest

unontenteza Common Tern

insukakude Arctic Tern

umachibini Whiskered Tern (br)

umachibini Whiskered Tern (non-br)

unochibi White-winged Tern

impisiyolwandle Brown Skua

DOVE & PIGEON

DOVE AND PIGEON GENERIC NAME. The more wide-spread generic terms for both doves and pigeons is **ihobhe**, with no apparent underlying meanings. Smaller groupings of doves and pigeons (see below) are referred to by the cluster names **ijuba** and **isikhombazane**. The name *ijuba*, in its plural form *amajuba*, is reflected in the name of Majuba Mountain near Newcastle, a battle site of the First Boer War, and a mountain long associated with the speckled pigeon, which falls under the generic *ijuba*.

ijuba ledolobha Rock Dove *Columba livia*

The generic name **ijuba** has been extended with *ledolobha* 'of the town/city'.

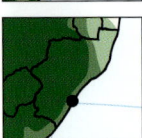

ivukuthu Speckled Pigeon *Columba guinea*

The Speckled Pigeon is one of the '*ijuba* doves'. The name **ivukuthu** accurately reflects the '*vuku vuku vukutu vu-u-u*' call of this bird.

An alternative name is **ijubantendele**, a name which effectively means the 'speckled pigeon'.

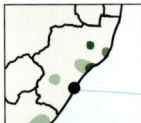

ijubelintamemhlophe Eastern Bronze-naped Pigeon *Columba delegorguei*

The generic name **ijuba** is extended here with *elintamemhlophe* 'white-necked'.

ivukuthu lehlathi African Olive Pigeon *Columba arquatrix*

The name **ivukuthu** has been extended here with *lehlathi* 'of the forest'.

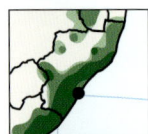

isagqukwe Lemon Dove *Columba larvata*

The name **isagqukwe** is well-known and widely-used for this bird. There is no apparent underlying meaning.

usamdokwe Ring-necked Dove *Streptopelia capicola*

The name **usamdokwe** is well-known and widely-used for this bird. The name appears to be derived from the prefix –*sa*– 'something like' with *umdokwe* 'porridge made from millet'. The bird's call is interpreted as "*umdokwe uvuthiwe, umdokwe uvuthiwe*" 'the porridge is ready' or 'the porridge millet is ripe'. The bird is a great nuisance in fields of ripe millet, hence this interpretation of the bird's call.

ijubelintamemhlophe Eastern Bronze-naped Pigeon ♀

ijuba ledolobha Rock Dove

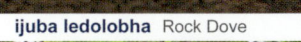

ivukuthu Speckled Pigeon

ijubelintamemhlophe Eastern Bronze-naped Pigeon ♂

ivukuthu lehlathi African Olive Pigeon

isagqukwe Lemon Dove

usamdokwe Ring-necked Dove

DOVE & PIGEON

ihobhelimehlabomvu Red-eyed Dove *Streptopelia semitorquata*
The generic name **ihobhe** is extended with *elimehlabomvu* 'which is red-eyed'.

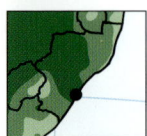

ukhonzane Laughing Dove *Streptopelia senegalensis*
The name **ukhonzane** is well-known and widely-used. It is almost certainly based on the verb *khonza* 'show respect', 'greet respectfully' in reference to the bird's nodding of the head and bowing.

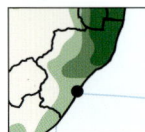

isikhombazane sehlanze, ingophozi Emerald-spotted Dove *Turtur chalcospilos*
The name **isikhombazane sehlanze** is well-known and widely-used. The extension – *sehlanze* means 'from the bushveld', and distinguishes this bird from the Tambourine Dove (see below).

An alternative name, less widely used, is **ingophozi**, with no obvious underlying meaning.

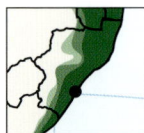

isikhombazane sehlathi Tambourine Dove *Turtur tympanistra*
The addition of –*sehlathi* of the forest distinguishes this dove from the Emerald-spotted Dove.

The alternative name **isibhelu**, with no obvious underlying meaning, is an old established name still commonly used in some areas.

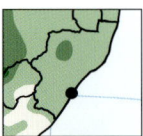

unkombose Namaqua Dove *Oena capensis*
The name **unkombose** is well-known and widely-used for this bird. It has been suggested that the name derives from *khomba* 'to point out' as the direct flight of this bird to water reveals the location of a water source. The alternative name **isikhombazane senkangala** 'isikhombazane of the open plain' is also widely-used for this bird.

ijubantondo African Green Pigeon *Treron calvus*
The well-known name **ijubantondo** for this bird supposedly means 'dove found in large numbers'. The name also occurs as **ijubantonto**.

ijubantondo African Green Pigeon feeding on fig fruits

ihobhelimehlabomvu Red-eyed Dove

ukhonzane Laughing Dove

isikhombazane sehlanze Emerald-spotted Dove

isikhombazane sehlathi Tambourine Dove ♂

unkombose Namaqua Dove ♂

ijubantondo African Green Pigeon

TURACO, GO-AWAY-BIRD & COUCAL

TURACO GENERIC NAME: igwalagwala. The old and well-established name **igwalagwala** is used as a generic name for three species of turaco found in KwaZulu-Natal. The feathers of this bird are used as head decorations for royalty among both the Zulu and the Swazi, and the name **igwalagwala** appears frequently in oral praise poetry, usually in the context of battle and spilt blood. The link between the red colour of blood and the red colour of the turaco feathers can be seen in the expression *ukumenza umuntu igwalagwala ekhanda* (lit. 'to do someone a 'gwalagwala' on the head', i.e. to strike someone such a blow on the head as to draw blood).

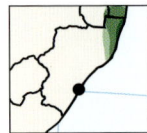

igwalagwala logu Livingstone's Turaco *Tauraco livingstonii*
The generic name **igwalagwala** is extended with *logu* 'of the coast'.

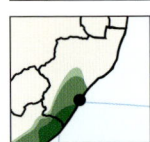

igwalagwaleliluhlaza Knysna Turaco *Tauraco corythaix*
The generic name **igwalagwala** has been extended with *eliluhlaza* 'which is green'.

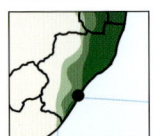

igwalagwala lehlanze Purple-crested Turaco *Tauraco porphyreolophus*
The generic name **igwalagwala** has been extended with *lehlanze* 'of the bushveld' for this bird to distinguish it from the other species of turaco.

umklewu Grey Go-away-bird *Corythaixoides concolor*
This onomatopoeic name is well-known for this bird and is similar to names in many of the other South African languages, including Afrikaans *kwêvoël*. There are a number of regional variations, including **umkliwu** and **umkluwe**. The English name – Grey Go-away-bird – incorporates the equally onomatopoeic *go-awaaaay*.

ufukwe, umgugwane Burchell's Coucal *Centropus burchellii*
Both **ufukwe** and **umgugwane** are well-known names for this bird. The name **ufukwe** has no obvious underlying meaning; **umgugwane** is possibly derived from *igugu* with the species-forming suffix *–ane*. The *igugu* is a large, black cockroach which scuttles into hiding when discovered, much as the coucal's hiding itself in dense reed-beds or thickets.

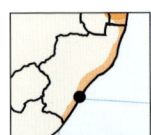

ufukwomnyama Black Coucal *Centropus grillii*
The generic name **ufukwe** is extended with *omnyama* 'which is black'.

igwalagwala lehlanze Purple-crested Turaco in date palm

ufukwomnyama Black coucal (juv)

igwalagwala logu Livingstone's Turaco

igwalagwaleliluhlaza Knysna Turaco

igwalagwala lehlanze Purple-crested Turaco

umklewu Grey Go-away-bird

ufukwe Burchell's Coucal

ufukwomnyama Black Coucal

MALKOHA & CUCKOO

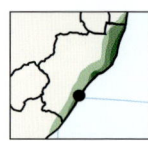

umcwicwicwi Green Malkoha *Ceuthmochares australis*
This coined name is onomatopoeia-based.

CUCKOO GENERIC NAME: unozalashiye:. This generic name for 'cuckoo' literally means 'that which lays [eggs] and leaves [them] behind'.

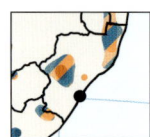

unozalashiyomabala Great Spotted Cuckoo *Clamator glandarius*
The generic name **unozalashiye** is extended with *omabala* 'which is spotted'.

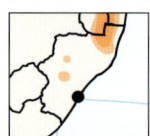

inkankemidwa Levaillant's Cuckoo *Clamator levaillanti*
This name is based on **inkanku** (the Jacobin Cuckoo), extended by *emidwa* 'which is striped'.

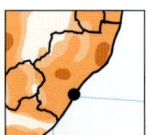

inkanku Jacobin Cuckoo *Clamator jacobinus*
This is a well-known and widely-used name for this bird. The call of this bird announces that spring is coming to an end and summer is nigh. The expression *inkanku isiwathethe amacimbi* 'the *inkanku* cuckoo has already taken the caterpillars' is a sign of mid-October.

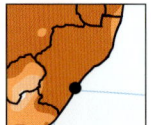

unonengekhanda Diederik Cuckoo *Chrysococcyx caprius*
The name **unonengekhanda** is in wide-spread oral usage, and is a 'verbalisation' of the call, with the literal meaning 'it became fat in the head'. The name is pronounced with a long and high note on the third syllable /ne/.

umazalashiye Klaas's Cuckoo *Chrysococcyx klaas*
This name is based on –*ma*– 'characteristically' with *zala* 'give birth' and *shiya* 'leave behind', i.e. the bird that lays eggs and then goes off leaving them behind. The name differs from the generic name only in the use of the prefix –*ma*– 'characteristically' instead of –*no*–.

unozalashiyomabala Great Spotted Cuckoo juv

umcwicwicwi Green Malkoha

unozalashiyomabala Great Spotted Cuckoo

inkankemidwa Levaillant's Cuckoo

inkanku Jacobin Cuckoo

unonengekhanda Diederik Cuckoo ♂

umazalashiye Klaas's Cuckoo ♂

CUCKOO

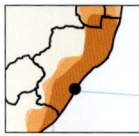

ubantwanyana African Emerald Cuckoo *Chrysococcyx cupreus*
This is a well-known and widely-used name, and a verbalisation of the bird's call. The bird supposedly sings "*Bantwanyana ningendi!*", i.e. 'Little children, don't get married!'

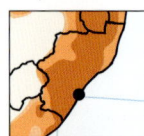

undodosibona Black Cuckoo *Cuculus clamosus*
This well-known and well-established name is a compound of *indoda* 'man' and *osibona* 'who sees us', and refers to the fact that this bird calls from dense vegetation, where people cannot see it, but it can see them.

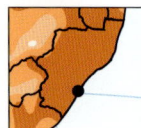

uphezukomkhono Red-chested Cuckoo *Cuculus solitarius*
This is a well-known name for this harbinger of spring. The name is onomatopoeic, as is the Afrikaans name *piet-my-vrou*, but it is also a verbalisation, a compound of *phezu[lu]* 'above', 'on top of' and *komkhono*, the locative form of *umkhono* 'upper arm'. There are a number of interpretations of this name, the most common and widely accepted being the interpretation of *umkhono* as 'shoulder' and the bird's song being a call to women to shoulder their hoes and go out to the fields to cultivate them.

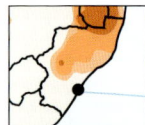

usakhukhuza African Cuckoo *Cuculus gularis*
This name is based on –*sa*– 'something like' and *unokhukhuza*, the name given to the Common Cuckoo (see below).

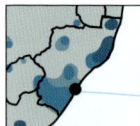

unokhukhuza Common Cuckoo *Cuculus canorus*
Like the English name of this bird (and names in many European languages), this has an onomatopoeic base. With the name-forming prefix –*no*– and the verb-forming suffix –*za* on either side of the onomatopoeic *khukhu* (< cuckoo) the name literally means 'the [bird] species that goes "Cuckoo!" '. However, it should be noted that this bird is silent when it is in southern Africa.

uphezukomkhono Red-chested Cuckoo chick being fed by Cape Wagtail host parent

ubantwanyana African Emerald Cuckoo ♂

ubantwanyana African Emerald Cuckoo ♀

undodosibona Black Cuckoo

uphezukomkhono Red-chested Cuckoo ♂

usakhukhuza African Cuckoo ♂

unokhukhuza Common Cuckoo ♂

OWL

OWL GENERIC NAME: isikhova. The use of **isikhova** as a generic name for owls is well-known and long-established. Like doves and pigeons, this cluster of birds is well-supplied with previously recorded species-specific names. The name **isikhovampondo** is a generic itself, referring to the three species of eagle owls. This previously recorded name is an extension of **isikhova**, formed by adding *[izi] mpondo* 'horns' in reference to the large ear tufts.

umzwelele Western Barn Owl *Tyto alba*
The name **umzwelele** is widely-used and has no obvious underlying meaning.

isikhova sotshani African Grass Owl *Tyto capensis*
This name extends the generic **isikhova** with *sotshani* 'of the grass'.

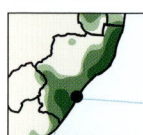

umabhengwane ♂, unobathekeli ♀ African Wood Owl *Strix woodfordii*
This is one of the few species of bird with separate names for the male and the female. The male is known as **umabhengwane** and the female as **unobathekeli**. While **umabhengwane** has no obvious underlying meaning, **unobathekeli** is derived from the noun *umthekeli*, which, among other meanings, simply means 'female'.

isikhova sexhaphozi Marsh Owl *Asio capensis*
The generic name **isikhova** has been extended with *sexhaphozi* 'of the marsh'.

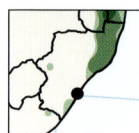

umadletshana, umloyi African Scops Owl *Otus senegalensis*
There are no previously recorded names for this owl. It is not clear whether the two names given here are coinages, or reflect names in general oral use. The name **umadletshana** is a diminutive form of *amadlebe* 'ears' and either means 'the bird with small ears' or 'the small bird with (distinctive) ears'. The name **umloyi** is presumably related to the usual meaning of this word, viz. 'witch, wizard, one who practises witchcraft', a reference to owls being night-birds and suspected of being familiars' 'helpers'. This belief extends in fact to all owls and such a belief is found all over Africa and in countries and cultures all over the world. There has been discussion about whether such beliefs and such names should continue to be recorded in publications lest they perpetuate beliefs which are counter to the conservation of owls generally.

umadletshana African Scops Owl camouflaged against tree trunk

umzwelele Western Barn Owl

isikhova sotshani African Grass Owl

umabhengwane African Wood Owl (rufous form)

umabhengwane African Wood Owl (brown form)

isikhova sexhaphozi Marsh Owl

umadletshana African Scops Owl

OWL & EAGLE-OWL

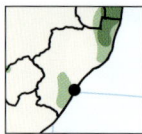
umandubulu Southern White-faced Owl *Ptilopsis granti*
The name **umandubulu**, with no obvious underlying meaning, has previously been used for the Pearl-spotted Owlet, but is now re-assigned to this species.

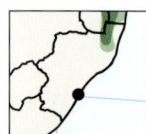
inkovana Pearl-spotted Owlet *Glaucidium perlatum*
This diminutive form of the generic name **isikhova** is well-known for this small owl.

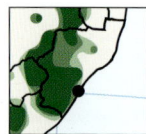
isikhovampondo Cape Eagle-Owl *Bubo capensis*
This well-known and long-established metaphor-based name for the eagle-owls uses the generic **isikhova** with the extension *[izi]mpondo* 'horns' – a reference to the prominent ear-tufts.

isikhovamponjwana Spotted Eagle-Owl *Bubo africanus*
The name **isikhovampondo** has had the diminutive suffix *–ana* added to it for this smaller version of the Cape Eagle-Owl.

isikhovanhlanzi Pel's Fishing Owl *Scotopelia peli*
The name **isikhovanhlanzi** is a compound of the generic **isikhova** and *inhlanzi* 'fish'.

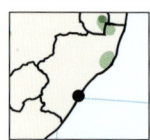
ifubesi Verreaux's Eagle-Owl *Bubo lacteus*
The name **ifubesi** is a well-known and long-established name for Verreaux's Eagle-Owl, with no apparent underlying meaning.

ifubesi Verreaux's Eagle-Owl chick feeding on frog

umandubulu Southern White-faced Owl

inkovana Pearl-spotted Owlet

isikhovampondo Cape Eagle-Owl

isikhovamponjwana Spotted Eagle-Owl

isikhovanhlanzi Pel's Fishing Owl

ifubesi Verreaux's Eagle-Owl

NIGHTJAR

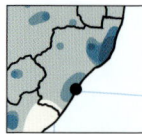

umawewe European Nightjar *Caprimulgus europaeus*
The name **umawewe** is widely used orally, and has no obvious underlying meaning.

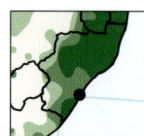

uzavolo Fiery-necked Nightjar *Caprimulgus pectoralis*
The name **uzavolo** is well-known and widely-used for this bird. Its call has been verbalised as *Zavolo! Zavolo! Sengel' abantabakho!* 'Zavolo! Zavolo! Go and milk [the cows] for your children!'. The underlying meaning of the Zulu verbalisation refers to the folk belief that these birds suck the teats of cattle and goats at night, a belief reflected in the common English name 'goatsucker' and the generic name *Caprimulgus*.

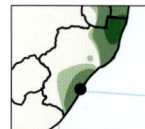

umalwelwe Square-tailed Nightjar *Caprimulgus fossii*
The name **umalwelwe** is similar to that of the European Nightjar, and like that name is widely used orally. It has no obvious underlying meaning.

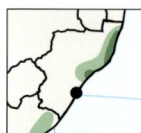

umhlohlongwane Swamp Nightjar *Caprimulgus natalensis*
This is a well-known name for this bird, with no known underlying meaning.

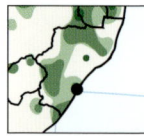

umatsheni Freckled Nightjar *Caprimulgus tristigma*
This name is derived directly from the locative noun *ematsheni* 'among the rocks', in reference to this bird's preferred behaviour during the day of roosting on bare or lichen-covered rocks where it is well-camouflaged.

uzavolo Fiery-necked Nightjar with chick hidden in leaves in daytime

umawewe European Nightjar

uzavolo Fiery-necked Nightjar

uzavolo Fiery-necked Nightjar (darker subspecies)

umalwelwe Square-tailed Nightjar

umhlohlongwane Swamp Nightjar

umatsheni Freckled Nightjar

SWIFT

SWIFT GENERIC NAME: ijiyankomo (often abbreviated to **ijankomo**). The fuller form **ijiyankomo** is derived from *jiya* 'be of an age' + *inkomo* 'head of cattle'. The primary meaning of the word 'ijiyankomo' is 'set of boys of a mature age' – i.e. old enough to herd cattle – and the secondary meaning 'common swift'. The association is unclear, but is assumed that because swifts are attracted to the flies and other insects disturbed by grazing cattle, they fly low and amongst the cattle. The name **ihlabankomo** 'what stabs the cattle' is also well-known and widely used for this reason.

A third name is **ihlolamvula** 'what predicts the rain', making the swift one of several so-called 'rainbirds' in Zulu culture (see African Black Swift below).

Yet another older name, less widely-known, is **uzalukelakude** 'what flies far away'.

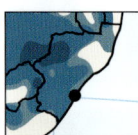

imvuliyeza Common Swift *Apus apus*
The name **imvuliyeza**, meaning 'the rain is coming', retains the Zulu cultural notion that swifts bring rain.

inhlolazulu Alpine Swift *Apus melba*
As with the Common Swift above, this name plays on the long-established name **ihlolamvula** 'what predicts the rain'. The name **inhlolazulu** is a compound of *hlola* 'predict' and *izulu* 'sky', 'weather', so a literal meaning is 'what predicts the weather'.

ihlolamvula African Black Swift *Apus barbatus*
As noted above, this name means 'what predicts the rain'.

ijiyankomelimlotha African Palm Swift *Cypsiurus parvus*
The generic name **ijiyankomo** is extended here with *elimlotha* 'which is ash-coloured'.

umakhalelilanga Little Swift *Apus affinis*
This name means 'that which characteristically cries at sun[rise] and sun[set]'.

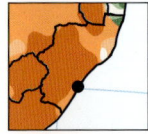

ijankomo Horus Swift *Apus horus*
The shortened form of the generic name **ijiyankomo** is used here unextended.

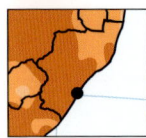

unonqane White-rumped Swift *Apus caffer*
The name **unonqane** is well-known and widely-used for this bird, and has no obvious underlying meaning.

imvuliyeza Common Swift

inhlolazulu Alpine Swift

ihlolamvula African Black Swift

ijiyankomelimlotha African Palm Swift

umakhalelilanga Little Swift

ijankomo Horus Swift

unonqane White-rumped Swift

MOUSEBIRD & ROLLER

indlazi Speckled Mousebird *Colius striatus*
A well-known and widely-used name for this bird, with no apparent underlying meaning.

umtshivovo, umjombo Red-faced Mousebird *Urocolius indicus*
The name **umtshivovo** is well-known and widely-used, and has been said to be onomatopoeic in origin.

An alternative name is **umjombo**, in current oral usage.

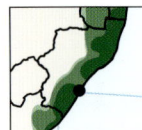

umjenenengu Narina Trogon *Apaloderma narina*
The name **umjenenengu** is well-known and widely-used. It has no obvious underlying meaning.

Also in use in some parts is the name **uzamkhuhlwini**, derived from *umkhuhlu* (Natal Mahogany, *Trichilia emetica*) which the bird is said to favour.

ROLLER GENERIC NAME: ifefe, with no underlying meaning.

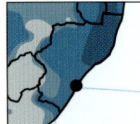

ifefeliluhlaza European Roller *Coracias garrulus*
The generic name **ifefe** has been extended with *eluhlaza* 'which is green'.

ifefemidwa Purple Roller *Coracias naevius*
The generic name **ifefe** has been extended with *emidwa* 'which is striped'.

ifefelihle Lilac-breasted Roller *Coracias caudata*
The generic name **ifefe** has been extended with *elihle* 'which is beautiful'.

umjenenengu Narina Trogon ♀ with caterpillar prey

indlazi Speckled Mousebird

umtshivovo Red-faced Mousebird

umjenenengu Narina Trogon ♂

ifefeliluhlaza European Roller

ifefemidwa Purple Roller

ifefelihle Lilac-breasted Roller

ROLLER & BEE-EATER

ifefelibomvu Broad-billed Roller *Eurystomus glaucurus*

The generic name **ifefe** has been extended with *ebomvu* 'which is red'.

KINGFISHER GENERIC NAME: indwazela. This name is based on the verb *dwazela* 'sit motionlessly while staring ahead'. This behaviour is a feature of kingfishers in general. The kingfisher cluster is one well-provided with previously recorded species-specific names, and the generic name **indwazela** has not been used once.

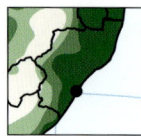

isiphikeleli Brown-hooded Kingfisher *Halcyon albiventris*

The name **isiphikeleli** is derived from the verb *phikelela* 'persist', 'persevere'. This is in reference to this bird persevering in pursuit of food from the same perch over and over again (see also Woodland Kingfisher).

Earlier alternative name: **unongozolo** (see Striped Kingfisher below).

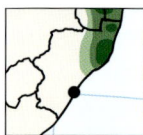

unongozolwane Striped Kingfisher *Halcyon chelicuti*

The name **unongozolo**, with no obvious underlying meaning, was earlier used for the Brown-hooded Kingfisher. This is the same name, but with the species-indicating suffix—*ane* added.

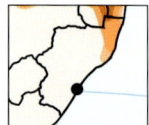

imbuyelelo Woodland Kingfisher *Halcyon senegalensis*

The name **imbuyelelo** is derived from the verb *buyelela* 'keep on returning', in reference to this bird always returning to exactly the same perch after swooping for prey (see also Brown-hooded Kingfisher above).

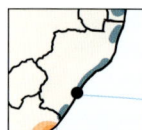

unonkalankala Mangrove Kingfisher *Halcyon senegaloides*

The name **unonkalankala** is based on the noun *inkalankala* 'crab' in reference to the bird's diet.

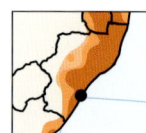

isikilothi African Pygmy Kingfisher *Ispidina picta*

The name **isikilothi** is well-known orally and is onomatopoeic in origin.

The alternative name **uzangume** is not commonly used for the bird today, but the meaning of 'one sitting staring vacantly into space' still applies to the medicinal use of this bird: parts are mixed with other ingredients to be made into a concoction to be given to a judge in a court case in which one fears to be found guilty. It is believed that such traditional medicine renders a judge immobile: instead of finding you guilty he will just sit and stare into space.

ifefelibomvu Broad-billed Roller

isiphikeleli Brown-hooded Kingfisher ♀

unongozolwane Striped Kingfisher

imbuyelelo Woodland Kingfisher

unonkalankala Mangrove Kingfisher

isikilothi African Pygmy Kingfisher

KINGFISHER & BEE-EATER

uzangozolo Malachite Kingfisher *Corythornis cristata*
This name is well-known and in current use. It is likely a regional variant of **unongozolo**, used previously for the Brown-hooded Kingfisher (see above).

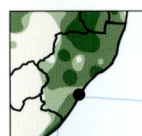

isixula Half-collared Kingfisher *Alcedo semitorquata*
This is a well-known and widely-used for this bird name, with no obvious underlying meaning.

isivuba Giant Kingfisher *Megaceryle maxima*
The name **isivuba** is well-known and widely-used. There is no clear underlying meaning.

ihlabahlabane Pied Kingfisher *Ceryle rudis*
The name **ihlabahlabane** is based on a reduplication of the verb *hlaba* 'stab', suggesting a continual 'stabbing' of the water surface with the beak when diving to catch fish. The commonly used suffix *–ane* has been added.

BEE-EATER GENERIC NAME: inkotha. The generic name **inkotha** is derived from the verb *khotha* 'scoop up'.

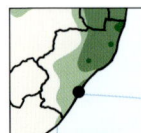

inkothana Little Bee-eater *Merops pusillus*
The generic name **inkotha** has been suffixed with the diminutive *–ana*, thus matching the English name of the bird. Another, earlier, name for this bird is **igundwana** 'mouse', a reference to the mouse-like tunnels this bird makes in river banks.

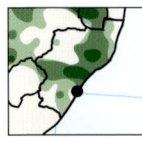

inkotha, uswenka White-fronted Bee-eater *Merops bullockoides*
The generic name **inkotha** is often used specifically to refer to this bird. In addition, a modern name in current use today is **uswenka**, derived from the English word *swanky*, a reference to the bright colours of this species.

isivuba Giant Kingfisher with tilapia prey ♂

uzangozolo Malachite Kingfisher

isixula Half-collared Kingfisher ♀

isivuba Giant Kingfisher ♀

ihlabahlabane Pied Kingfisher ♀

inkothana Little Bee-eater

inkotha White-fronted Bee-eater

BEE-EATER, HOOPOE & SCIMITARBILL

indlanyosi Blue-cheeked Bee-eater *Merops persicus*
This name is formed from the verb *dla* 'to eat' and the noun *inyosi* 'bee'.

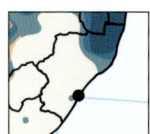

inkothanyosi European Bee-eater *Merops apiaster*
This name is a compound of *khotha* 'to catch' and the noun *inyosi* 'bee'.

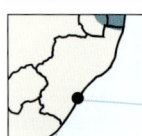

inkothenkulu Southern Carmine Bee-eater *Merops nubicoides*
The generic name **inkotha** has been extended with *enkulu* 'which is big'.

umzolozolo, umambathingubo African Hoopoe *Upupa africana*
The name **umzolozolo** is well-known and widely-used. It has no obvious underlying meaning. The name **umambathingubo**, meaning 'what characteristically wears a colourful blanket', is an earlier name which has been retained.

inhlekabafazi, unukani Green Wood Hoopoe *Phoeniculus purpureus*
The name **inhlekabafazi** is usually interpreted as meaning 'the laughter of women', because the sound of a group of these birds all cackling together sounds like a group of women laughing. It is derived from *hleka* 'laugh' and *abafazi* 'women'. Another commonly used name for this bird is **unukani**, meaning 'smells of what?', a reference to the odorous nests of these birds. A lesser-known alternative name is **ugoligoli**, used only in the north-eastern regions of KwaZulu-Natal. The name has no obvious underlying meaning.

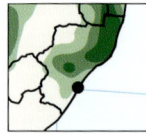

unosungulo Common Scimitarbill *Phoeniculus cyanomelas*
The name **unosungulo** is based on the noun *usungulo* 'long, curved needle', a reference to the 'scimitar bill' of this bird.

indlanyosi Blue-cheeked Bee-eater with dragonfly prey

indlanyosi Blue-cheeked Bee-eater

inkothanyosi European Bee-eater

inkothenkulu Southern Carmine Bee-eater

umzolozolo African Hoopoe

inhlekabafazi Green Wood Hoopoe

unosungulo Common Scimitarbill

HORNBILL & GROUND HORNBILL

HORNBILL NAME There is no generic name for hornbills as they are not perceived as a 'cluster' of related birds in Zulu folk taxonomy.

umkhololwane Crowned Hornbill *Lophoceros alboterminatus*

The name **umkhololwane** is an adaptation of the name **umkholwane** for the Southern Red-billed Hornbill (see below).

umkholwane Southern Red-billed Hornbill *Tockus erythrorhynchus*

This is an old name in wide-spread use for this bird. It is likely formed from the species-forming suffix –*ane* and a very old root –*kholo*, with no obvious meaning.

umkholwanomlotha African Grey Hornbill *Lophoceros nasutus*

The name **umkholwane** (see above) has been extended with *omlotha* 'which is ash-coloured' for this species.

umkholompunga, uzazu Southern Yellow-billed Hornbill *Tockus leucomelas*

The root –*kholo* has been extended with *mpunga* 'grey' to form **umkholompunga**. A more modern alternative name is **uzazu**, derived from the name of the popular yellow-billed hornbill character in Disney's film *The Lion King*.

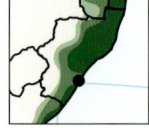

imemela, injubalukhalo Trumpeter Hornbill *Bycanistes bucinator*

The name **imemela** is onomatopoeic, and refers to the bird crying *me-e-e-h me-e-eh*. The alternate name **injubalukhalo** apparently also refers to the bird's nasal, wailing call but the structure of the word is not possible to determine. The Zulu verb *khala* 'wail, cry, lament' appears to be part of the name.

The earlier alternate name **ikhunatha** (from the verb *khunatha* 'be sulky') appears to be no longer in popular use.

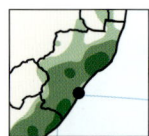

insingizi, ingududu Southern Ground Hornbill *Bucorvus leadbeateri*

The name **insingizi**, with no obvious underlying meaning, is an old name with wide-spread usage. Equally well-known is the name **ingududu**, but as this is an onomatopoeic name, it is usually only used when the birds are heard calling. When these birds are duetting, the female is believed to be saying "*Sengiyemuka! Sengiyemuka!*" 'I am leaving now' to which the male answers "*Hamba! Kad' usho! Hamba! Kad' usho!*" 'Go then! You have been saying so for long enough.'

insingizi Southern Ground Hornbill can kill and devour large snakes

umkhololwane Crowned Hornbill

umkholwane Southern Red-billed Hornbill

umkholwanomlotha African Grey Hornbill ♂

umkholompunga Southern Yellow-billed Hornbill

imemela Trumpeter Hornbill

insingizi Southern Ground Hornbill

BARBET & TINKERBIRD

BARBET NAME There is no generic name for barbets in Zulu as they do not constitute a recognisable cluster of birds in Zulu folk taxonomy.

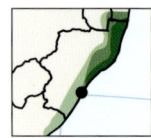

intunjana White-eared Barbet *Stactolaema leucotis*
The name is a diminutive of *intunja* 'small aperture' and could be related to the fact that the nesting bird sits in an excavated hole looking out through a narrow aperture.

unongoyana Green Barbet *Stactolaema olivacea*
The name **unongoyana** is formed from Ngoye, the name of the forest in KwaZulu-Natal where this bird is specifically found. The word has been prefixed with –*no*– and suffixed with –*ana* giving the meaning 'the little [bird] from the Ngoye'.

usiqhovana Crested Barbet *Trachyphonus vaillantii*
The name **usiqhovana** is derived from the word *isiqhova* 'crest' and the diminutive suffix *ana*, thus meaning '[the bird with the] little crest'.

usibagwebe, isinqonqotho, isandondondwane Black-collared Barbet *Lybius torquatus*
This bird has a number of currently used Zulu names, and all refer to the sounds made by the bird. The three given here all refer to the knocking or tapping sounds made by the bird: **isinqonqotho** and **isandondondwane** are both onomatopoeic (standing at a doorway and saying "Nqo! Nqo!" is the Zulu way of knocking on a door), while **usibagwebe** means 'we stab them' and refers to the bird stabbing its beak into tree surfaces in search of food.

Another name in common use is **isikhuhlukhuhlu (isikhulukhulu** in some regions) which is onomatopoeic for the bird's duetting call.

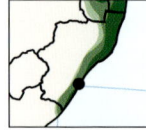

isipopopo Yellow-rumped Tinkerbird *Pogoniulus bilineatus*
The name **isipopopo** is onomatopoeic, referring to the bird's *pot-pot-pot* call.

unkovuka Red-fronted Tinkerbird *Pogoniulus pusillus*
The name **unkovuka** is known only from the northern-most parts of KwaZulu-Natal. It has no obvious underlying meaning. An earlier name for this bird is **unogandilanga**, meaning '[the bird that] pounds all day long', and this name can be used as well.

intunjana White-eared Barbet feeding on pawpaw

intunjana White-eared Barbet in nest cavity

intunjana White-eared Barbet

unongoyana Green Barbet

usiqhovana Crested Barbet

usibagwebe Black-collared Barbet

isipopopo Yellow-rumped Tinkerbird

unkovuka Red-fronted Tinkerbird

BARBET, HONEYGUIDE & WRYNECK

unomunga Acacia Pied Barbet *Lybius leucomelas*
This name is derived from *umunga*, a generic name for a number of acacia trees, notably *Vachellia tortilis*.

HONEYGUIDE GENERIC NAME: ingede A number of names have been previously recorded for honeyguides in general, including **ingede**, **ihlava**, **inhlavebizelayo** and **unomtsheketshe**, all of which are used for the different species of honeyguide found in KwaZulu-Natal. The name **ingede** is given here as a generic name for honeyguides as the most commonly used name for this group of birds. The behaviour of the honeyguide is well known in traditional societies all over Africa, and everywhere there is the general belief that when the bird leads someone to a beehive, it is appropriate to share some of the honey with it. This belief is reflected in two Zulu proverbs:

Inhlav' iyabekelwa (The honeyguide is given something), meaning you should always show gratitude towards someone who has helped you.

Ungayishay' ingede ngoju (Do not strike the honeybird with honey.) This equates to the English 'Do not bite the hand that feeds you'.

unomtsheketshe Brown-backed Honeybird *Prodotiscus regulus*
This name is one of the several names used for honeyguides. The name is based on the noun *itsheketshe* 'ant', a reference to the insectivorous diet of honeyguides.

ingedana Lesser Honeyguide *Indicator minor*
The name **ingede** has been suffixed with the diminutive *–ana* to form this name, matching both 'lesser' in the vernacular name and 'minor' in the scientific name.

inhlavana Scaly-throated Honeyguide *Indicator variegatus*
The name **inhlava** has been suffixed with the diminutive *–ana* to form this name.

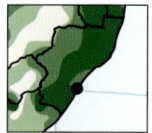

inhlavebizelayo, ingede Greater Honeyguide *Indicator indicator*
Both **inhlavebizelayo** and **ingede** are commonly used for this honeyguide. The name **inhlavebizelayo** is a compound of *inhlava* and *ebizelayo* 'which is calling to [someone]' from the verb *biza* 'call out'. This specific species is the most well-known for calling and leading people to beehives.

unongilobomvu Red-throated Wryneck *Jynx ruficollis*
This name is derived from the prefix *–no–* with *ngilo* 'throat' and *bomvu* 'red'.

unomunga Acacia Pied Barbet

unomtsheketshe Brown-backed Honeybird

ingedana Lesser Honeyguide

inhlavana Scaly-throated Honeyguide

inhlavebizelayo Greater Honeyguide ♂

unongilobomvu Red-throated Wryneck

WOODPECKER

WOODPECKER GENERIC NAME: isigqobhamithi. The generic name **isigqobhamithi** is a compound of *gqobha* 'peck, as a hen or woodpecker' and *imithi* 'trees'. An earlier recorded form of the name is **isiqophamuthi**, from *qopha* 'peck holes in' and *umuthi* 'tree'.

isigqobhamithi, usagwebe Golden-tailed Woodpecker *Campethera abingoni*

The generic name **isigqobhamithi** is used for this species of woodpecker, as well as the name **usagwebe**, an abbreviated form of **usibagweba** 'we stab them' (see entry for Black-collared Barbet above).

isigqobhamithi saseningizimu Knysna Woodpecker *Campethera notata*

The generic name **isigqobhamithi** has been extended with *saseningizimu* 'from the south'.

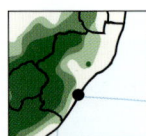

umnqangqandolo Ground Woodpecker *Geocolaptes olivaceus*

This is a well-known and widely-used name for this bird, with no clear underlying meaning.

inqondanqonda Cardinal Woodpecker *Dendropicos fuscescens*

The name **inqondanqonda**, like so many names for the woodpeckers and barbets, refers to the knocking sound made by this bird.

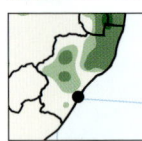

isigqobhamithintshebe Bearded Woodpecker *Dendropicos namaquus*

This name is formed by compounding **isigqobhamithi** with *intshebe* 'beard'. (Cf. **ukhozilwentshebe** for the Bearded Vulture.)

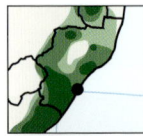

isigqobhamithesiluhlaza Olive Woodpecker *Dendropicos griseocephalus*

The generic name **isigqobhamithi** has been extended with *esiluhlaza* 'which is green'.

isigqobhamithesiluhlaza Olive Woodpecker ♂ with prey at nest hole

isigqobhamithi Golden-tailed Woodpecker ♂

isigqobhamithi saseningizimu Knysna Woodpecker ♂

umnqangqandolo Ground Woodpecker

inqondanqonda Cardinal Woodpecker ♂

isigqobhamithintshebe Bearded Woodpecker ♂

isigqobhamithesiluhlaza Olive Woodpecker ♂

PARAKEET, PARROT, BROADBILL & BATIS

unocu Rose-ringed Parakeet *Psittacula krameri*

The name **unocu** is formed by prefixing –*no*– to the word *ucu*, referring to the bead necklace a Zulu girl gives to a boy if she accepts him as a lover.

isikhwenene Cape Parrot *Poicephalus robustus*

The name **isikhwenene**, with no obvious underlying meaning, is well-known and widely-used for this species.

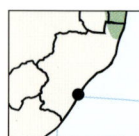

isikhwenenesikhandansundu Brown-headed Parrot *Poicephalus cryptoxanthus*

The name **isikhwenene** has been extended with *esikhandansundu* 'brown-headed' for this bird.

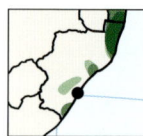

umasikulufu African Broadbill *Smithornis capensis*

The name **umasikhulufu** has as its base –*ma*– 'characteristically' and the adopted noun *isikulufu* (from Eng. *screw*). The bird has a horizontal circular display flight which suggests the turning of a screw.

umashiyabomvu Black-throated Wattle-eye *Platysteira peltata*

The name **umashiyabomvu** is a compound of *[a]mashiya* 'eyebrows' and *bomvu* 'red'.

BATIS GENERIC NAME: umnqube The name **umnqube**, with no obvious underlying meaning, is one of several names used for the different Batis species found in KwaZulu-Natal.

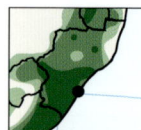

imbove, udokotela Cape Batis *Batis capensis*

The name **imbove**, with no obvious underlying meaning, is in wide-spread current use for this bird.

Another name, also in current use, is the name **udokotela**, from the English word 'doctor', a reference to the marking on the neck of the bird, which resemble the skin bag (*umhlanti*) worn by a traditional doctor (*inyanga*) to carry his medicines, (see also **incwaba** for the Chin-spot Batis below).

umashiyabomvu Black-throated Wattle-eye ♀

imbove Cape Batis ♀

unocu Rose-ringed Parakeet

isikhwenene Cape Parrot

isikhwenenesikhandansundu Brown-headed Parrot

umasikulufu African Broadbill ♂

umashiyabomvu Black-throated Wattle-eye ♂

imbove Cape Batis ♂

BATIS, HELMETSHRIKE & BUSHSHRIKE

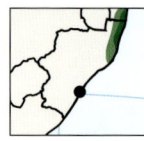

umnqube wogu Woodwards' Batis *Batis fratrum*
The generic name **umnqube** is extended with *wogu* 'of the coast'.

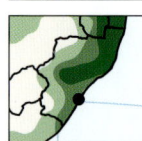

incwaba Chinspot Batis *Batis molitor*
The name **incwaba** is a regional variant of the word *incweba* (also as *ingqaba* and *umhlanti*), which means 'a small skin-bag containing medicines or charms and worn with others on a string around the neck'. See also the entry for the Cape Batis above and its name **udokotela**.

HELMETSHRIKE GENERIC NAME: impevu or iphemvu Both the regional variants **impevu** and **iphemvu** are in common use for the helmetshrikes.

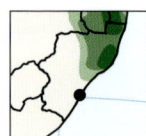

abayeni, ithimbakazane White-crested Helmetshrike *Prionops plumatus*
The name **abayeni** is a plural form of the noun *umyeni* 'husband'. The name **ithimbakazane** is linked to a number of similar-sounding Zulu words (e.g. *ithimbana*, *umthimbazana*) which all relate to groups of young girls, particularly as members of a bridal party. Clearly what is being referred to in both the names of this bird is the way the birds congregate in small groups that move through the trees chattering to one another just as the lively groups which separately accompany the bride and the groom do when on their way to a Zulu wedding.

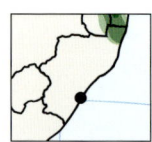

impevelimehlabomvu Retz's Helmetshrike *Prionops retzii*
The generic name **impevu** has been extended with *elimehlabomvu* 'red-eyed'.

BUSHSHRIKE GENERIC NAME The name **igqumusha** has long been used as a generic name for bushshrikes. It is derived from the ideophone *gqúmu* 'of standing out prominently' and refers to the way in which these birds proudly show themselves off. A well-known Zulu proverb is *insimba yasulela ngegqumusha* (literally: 'the genet wiped himself on the bushshrike'), and meaning 'people who are caught in misdemeanours often try to put the blame on others'. The basis of the proverb is an equally well-known Zulu *inganekwane* 'folktale', which can be summarised as follows:

The king of the animals and birds has obtained something valuable and desirable, for example a quantity of honey from a beehive. He places a guard over it and warns all the other animals and birds not to go near it. In the night, the genet sneaks off and by tricking the guards, gets access to the honey, of which he eats a quantity. Some honey sticks to his fur and he realises that when the theft is discovered he will be identified as the thief and punished. He then wipes himself clean on the sleeping bushshrike. In the morning there is a great hullabaloo when the theft is discovered and the king orders an investigation. At this point, the genet suggests checking the fur or plumage of all the animals and birds. Naturally the bushshrike is soon discovered with honey smeared all over him, and he is taken away and punished. The genet gets away scot-free.

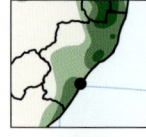

usipoki Grey-headed Bushshrike *Malaconotus blanchoti*
The name **usipoki** is derived directly from the Zulu noun *isipoki* 'ghost', itself adopted from either English or Afrikaans *spook*. This is a reference to the bird's ghostly, mournful, eerie whistle. Afrikaans uses the same metaphor in *spookvoël* 'ghost bird'.

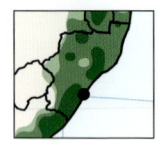

umabhashinhlayelohlaza Olive Bushshrike *Chlorophoneus olivaceus*
This name uses the onomatopoeic name **umabhashinhlayela** used for the Orange-breasted Bushshrike (see entry below) and extends it with *ohlaza* 'which is green'.

umnqube wogu Woodwards' Batis ♂

incwaba Chinspot Batis ♂

abayeni White-crested Helmetshrike

impevelimehlabomvu Retz's Helmetshrike

usipoki Grey-headed Bushshrike

umabhashinhlayelohlaza Olive Bushshrike

BUSHSHRIKE, BOKMAKIERIE, BRUBRU, PUFFBACK AND BOUBOU

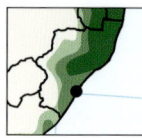

umabhashinhlayela Orange-breasted Bushshrike *Chlorophoneus sulfureopectus*

The name **umabhashinhlayela** is an onomatopoeic name and should be pronounced with five quick syllables *u-ma-bha-shi-nhla* and then the sixth syllable *ye* lengthened *(ye-e-e-e)* and on a higher pitch, with the final syllable barely audible.

The name **uhlaza** has also regularly been used for this bird, but also for various other species of shrikes.

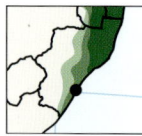

ingongoni, isangulube Gorgeous Bushshrike *Telephorus viridis*

Both these names refer to the bird's call. The name **ingongoni** is onomatopoeic as is the Afrikaans name *konkoit* for this bird. It is a long-established name.

The name **isangulube** is based on a simile: it means 'like a pig' and refers to the grunting *oink oink* call.

inkovu Bokmakierie *Telephorus zeylonus*

This is a well-known and widely used name for this bird, with no obvious underlying meaning.

usacingo Brubru *Nilaus afer*

The name **usacingo** is based on the prefix *–sa–* 'something like' and the noun *ucingo* 'wire', 'telephone'. The name is a reference to the repetitive purring trill of the bird.

isicivo Black-backed Puffback *Dryoscopus cubla*

The name **isicivo** is onomatopoeic, and should be pronounced with emphasis on the final syllable: *i-si-ci-vó*.

The onomatopoeic name **ibhoboni** is also well-known, but should be used with caution as it is also widely used for the Southern Boubou.

ibhoboni Southern Boubou *Laniarius ferrugineus*

The onomatopoeic name **ibhoboni** is commonly used to refer to this species.

The name **igqumusha** has and is also used for this bird, as well as for a number of other species of shrike.

isicivo Black-backed Puffback ♂

umabhashinhlayela Orange-breasted Bushshrike

ingongoni Gorgeous Bushshrike ♂

inkovu Bokmakierie

usacingo Brubru

isicivo Black-backed Puffback ♀

ibhoboni Southern Boubou

TCHAGRA, CUCKOOSHRIKE & SHRIKE

umngquphane Black-crowned Tchagra *Tchagra senegalus*
The name **umnquphane**, with no obvious underlying meaning, is widely-used for this bird.

isikhwayimba Brown-crowned Tchagra *Tchagra australis*
The name **isikhwayimba,** with an underlying meaning 'broad, flat, stiff object', may refer to the bird's flight with broad-spread wings and tail. The name has also been used in its diminutive form **isikhwayimbana** for this bird.

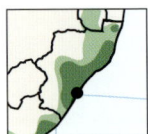

ushowe Southern Tchagra *Tchagra tchagra*
This onomatopoeic name comes from the very northern part of KwaZulu-Natal where it is widely used as a generic for all tchagras.

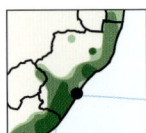

iklebedwane Grey Cuckooshrike *Coracina caesia*
The name **uklebedwane** is well-known and widely-used, with no clear underlying meaning.

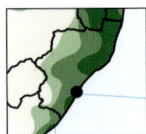

inhlangu Black Cuckooshrike *Campephaga flava*
A well-known name that is widely-used. The same word is used to refer to the reedbuck.

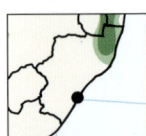

umqonqotho; ikhonqwelo Magpie Shrike *Urolestes melanoleucus*
The name **umqonqotho**, usually a reference to knocking sounds, is probably a reference here to the group display of these birds, where they bob and bow to each other while calling. The alternative name **ikhonqwelo** is a well-known and long-established name for this bird, of no apparent underlying meaning.

inhlangu Black Cuckooshrike ♀

142

umngquphane Black-crowned Tchagra

isikhwayimba Brown-crowned Tchagra

ushowe Southern Tchagra

iklebedwane Grey Cuckooshrike

inhlangu Black Cuckooshrike ♂

umqonqotho Magpie Shrike

SHRIKE, ORIOLE & ROCKJUMPER

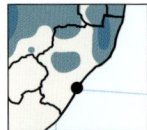

unolunga Lesser Grey Shrike *Lanius minor*
This name is an adaptation, via the prefix *–no–*, of **ilunga**, a name for the Southern Fiscal.

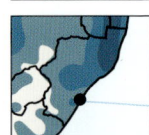

umathithibala Red-backed Shrike *Lanius collurio*
The name **umathithibala** is also the name of the aloe-like succulent *Haworthia limifolia*, which is used as a protective charm, being planted around the edges of a homestead to repel evil. The name has been applied to this shrike as the female has markings very similar to those on the leaves of the plant.

iqola, ilunga Southern Fiscal *Lanius collaris*
Both these names are equally well-known and widely-used. Both names are also applied to cattle with similar colours, **ilunga** referring to a black or brown beast with white stripes across stomach or legs, and *iqola* to a black beast with white markings across the back and the sides. The word *ilunga* is used in some bird names to indicate pied coloration (see **inkenkanelunga** for the Sacred Ibis and **umvemvelunga** for the African Pied Wagtail).

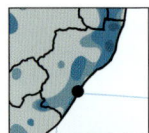

impofana Eurasian Golden Oriole *Oriolus oriolus*
The black and yellow plumage of this bird calls to mind the similar colours worn by the players of the Kaizer Chiefs Football Club, whose nickname is Mpofana.

umqoqongo, usibó Black-headed Oriole *Oriolus larvatus*
This bird has a number of previously recorded names. The names that are in current use, both being onomatopoeic, are *umqoqongo* and *usibó*, the latter pronounced with the stress on the last syllable: *u-si-bó*.

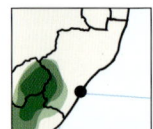

unogxumetsheni Drakensberg Rockjumper *Chaetops aurantius*
This name is based on the prefix *–no–* with *gxuma* 'jump' and *etsheni* 'on the rock'.

unogxumetsheni Drakensberg Rockjumper ♀

umathithibala Red-backed Shrike ♀

unolunga Lesser Grey Shrike

umathithibala Red-backed Shrike ♂

iqola Southern Fiscal

impofana Eurasian Golden Oriole

Adam Riley - Rockjumper Birding

umqoqongo Black-headed Oriole

unogxumetsheni Drakensberg Rockjumper ♂

DRONGO & FLYCATCHER

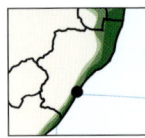

uhlakayiya, intengwana Southern Square-tailed Drongo *Dicrurus ludwigii*

The name **uhlakayiya**, an onomatopoeic name which refers to the variety of strident sounds this bird makes, is used in the very northernmost parts of KwaZulu-Natal. In central Zululand, and south of the uThukela River, the name **intengwana** is more widely used. It is a diminutive form of **intengu**, the name of the following species.

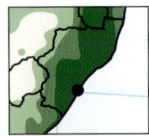

intengu Fork-tailed Drongo *Dicrurus adsimilis*

The name **intengu** is an old name still in wide-spread use for this bird. It has no apparent underlying meaning.

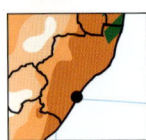

inzwece, u(lu)ve African Paradise Flycatcher *Terpsiphone viridis*

The name **inzwece** is onomatopoeic and in wide-spread use. The name **uve** (occurring also as **uluve**) is equally well-known, and has no apparent underlying meaning.

During the time of the Zulu kings in the 19th century, the long tail feathers of the male Paradise Flycatcher were the only feathers used exclusively by the kings. This practice seems to have fallen into abeyance.

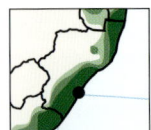

uqholwane Blue-mantled Crested Flycatcher *Trochocerus cyanomelas*

This name is based on the noun *iqholo*, which refers to a bunch of feathers worn on top of the head by young men in traditional dress. The species-indicating suffix –*ane* has been added to this to form **uqholwane**.

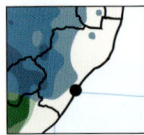

uqholompunga Fairy Flycatcher *Stenostira scita*

This name uses the same base *iqholo* as the name of the previous species, but instead of suffixing – *ane*, the base is extended with –*mpunga* 'grey'.

uqholwane Blue-mantled Crested Flycatcher ♀

uhlakayiya Southern Square-tailed Drongo

intengu Fork-tailed Drongo

inzwece African Paradise Flycatcher ♂

inzwece African Paradise Flycatcher ♂

uqholwane Blue-mantled Crested Flycatcher ♂

uqholompunga Fairy Flycatcher

CROW, RAVEN & TIT

CORVID GENERIC NAMES: In Zulu and in other African languages, the names of corvids tends to include the onomatopoeic syllables *kwa*, *gwa* or *kra*. This phenomenon is in fact found in names for corvids the world over and is seen in English *crow* and Afrikaans *kraai*. It is seen in the first syllable of the Zulu generic name **igwababa**.

igwababa ledolobha House Crow *Corvus splendens*
The generic name **igwababa** has been extended with *ledolobha* 'of the town/city' for this alien invasive species.

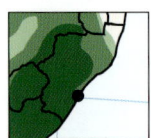

igwababakazi Cape Crow *Corvus capensis*
The generic name **igwababa** has been suffixed with –*kazi*, signifying 'larger' here, as this bird is larger than the House Crow.

igwababa Pied Crow *Corvus albus*
The Pied Crow carries the generic name **igwababa** without any extensions, adaptations or qualifications.

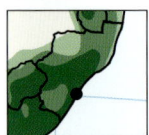

iwabayi White-necked Raven *Corvus albicollis*
The White-necked Raven has a number of names, of which **iwabayi**, of no apparent underlying meaning, is the most commonly used. The names **uhlungulu** and **ihubhulu**, also of no clear underlying meaning, are still found in some regions.

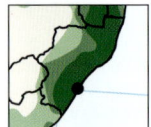

isicukujeje Southern Black Tit *Melaniparus niger*
There appears to be no apparent underlying meaning for this known name for the Southern Black Tit, although the last element 'jeje' clearly reflects the *je je je* call of this bird.

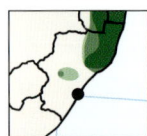

iklosi Grey Penduline Tit *Anthoscopus caroli*
This name is onomatopoeic, reflecting the rasping nature of this bird's call.

igwababa Pied Crow nest on telephone pole

igwababa ledolobha House Crow

igwababakazi Cape Crow

igwababa Pied Crow

iwabayi White-necked Raven

isicukujeje Southern Black Tit

iklosi Grey Penduline Tit

149

LARK

LARK GENERIC NAME: ucilo The lark cluster, comprising several species difficult to distinguish from one another, share a number of names. These names are assigned below to different individual species. The name **ucilo**, with no obvious underlying meaning, is used as the generic name for larks.

The **ucilo** plays quite an important role in Zulu folk literature. In the tale "The King of the Birds", the lark is the one that hides among the feathers on the eagle's back when the birds compete to see who can fly the highest. When the eagle flies higher than all the other birds, the lark climbs out from its hiding place and flies just a little higher, thus claiming the title of 'king of the birds'.

The **ucilo** also appears in two proverbs and/or expressions:

Ucilo uyilahlile intethe kubani (lit. 'the lark has dropped a grasshopper on So-and-so') = It is all up with So-and-so, or So-and-so is expected to die soon (it is believed that once an ucilo lark has caught a grasshopper, you will have to kill the bird to make it let go of its prey).
Ucilo akafi izidubuli (lit. 'the ucilo lark does not die from his blows') = don't mind hard knocks in life, you will get over them.

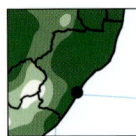

untilontilo, uqaqashe Rufous-naped Lark *Mirafra africana*

The name **untilontilo**, with no obvious underlying meaning, is a name in current use. The name **uqaqashe**, also with no apparent underlying meaning, is an alternative.

The Rufous-naped Lark and the Flappet Lark (see below) are considered indicators of luck in traditional Zulu culture: if one appears in front of you on the path and then suddenly flies off to one side, this is considered bad luck and something unfortunate will happen to you soon. If on the other hand it continues to run in front of you, then you will have good luck. If you are part of a group of courting youths, this is interpreted as meaning you will soon meet up with a group of lovely girls who will look on you with favour. This belief relates to the underlying meaning of the name **ungqangendlela** (see below).

ungqangendlela Flappet Lark *Mirafra rufocinnamomea*

This name for the Flappet Lark means 'what runs straight along the path' and is derived from the ideophone *ngqá* 'of straight direction' and *ngendlela* 'along the path', and refers to one of the indicators of good luck that larks in general are known for.

unqothi Sabota Lark *Calendulauda sabota*

The name **unqothi** is in current oral usage. It has no apparent underlying meaning.

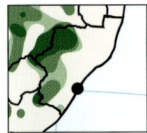

unonzwili Eastern Long-billed Lark *Certhilauda semitorquata*

The name **unonzwili** has been derived from the prefix *–no–* and an abbreviated form of the name **umzwilili** 'canary' in reference to the melodious song of this bird.

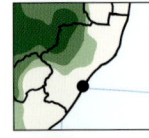

isanqunzi Spike-heeled Lark *Chersomanes albofasciata*

This name, previously unrecorded, appears to be based on *–sa–* 'something like' and '*inqunzi*', an untraceable noun. The derivation is obscure.

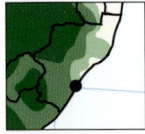

umntoli Red-capped Lark *Calandrella cinerea*

A well-known and long-established name for this bird, it has no apparent underlying meaning.

untilontilo Rufous-naped Lark

ungqangendlela Flappet Lark

unqothi Sabota Lark

unonzwili Eastern Long-billed Lark

isanqunzi Spike-heeled Lark

umntoli Red-capped Lark

LARK, SPARROW-LARK, NICATOR, BULBUL & GREENBUL

unongqwashi Pink-billed lark *Spizocorys conirostris*
One of the several names for the lark cluster has been assigned to this species. It has no apparent underlying meaning.

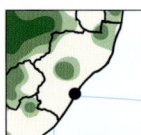

intakajolwane Chestnut-backed Sparrow-Lark *Eremopterix leucotis*
The name **intakajolwane** is a compound of **intaka** 'finch' and **ujolwane** 'sparrow'.

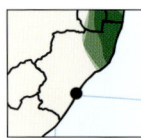

umalusinkomo Eastern Nicator *Nicator gularis*
This name has the underlying meaning 'the one that characteristically herds cattle'. This species is known to accompany feeding animals, both indigenous and domestic types, in bushy areas, waiting until the animals disturb the bird's insect prey.

iphothwe Dark-capped Bulbul *Pycnonotus tricolor*
This is a well-known and long-established name for this bird.

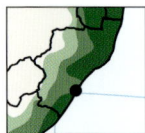

uwili, umashwili Sombre Greenbul *Andropadus importunus*
The name **uwili** (or *iwili*) is well-known and widely-used. This name is onomatopoeic (cf. the Afrikaans name *Gewone Willie*). The name **umashwili** is also in general usage.

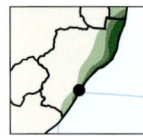

ibhada, unanaza Yellow-bellied Greenbul *Chlorocichla flaviventris*
The name **ibhada** is well-known for this bird, with no apparent underlying meaning. The name **unanaza** is also in general usage and refers to the nasal quality of the bird's call.

intakajolwane Chestnut-backed Sparrow-Lark ♀ (right) feeding juv

unongqwashi Pink-billed lark

intakajolwane Chestnut-backed Sparrow-Lark ♂

umalusinkomo Eastern Nicator

iphothwe Dark-capped Bulbul

uwili Sombre Greenbul

ibhada Yellow-bellied Greenbul

GREENBUL, BROWNBUL, SWALLOW, SAW-WING & MARTIN

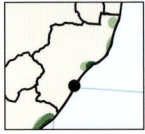

umashwilomidwa Yellow-streaked Greenbul *Phyllastrepus flavostriatus*

The name **umashwili** (see entry for Sombre Bulbul above) has been extended with *omidwa* 'which is striped or streaked'.

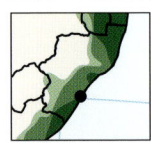

igedezi Terrestrial Brownbul *Phyllastrephus terrestris*

The name **igedezi** is derived from the verb *gedeza*, meaning 'talk incessantly, make a din while talking'. When disturbed, a group of these makes a series of harsh chattering, churring sounds.

SWALLOW AND MARTIN GENERIC NAMES: The name **inkonjane** as a generic for swallows is well-known. The **inkonjane** appears in the *izithakazelo* (clan praises) of a number of Zulu clans (including the Duma, Khumalo, Mlotshwa, Mnguni, Mthombeni and Ntuli clans) in slightly different variations of the following lines:

Ulwandle kaluwelwa (the sea is not crossed)

Luwelwa izinkonjane (it is [only] crossed by the swallows)

Zona zindiza phezulu (those that fly above)

These *izithakazelo*, which are a mixture of the names of ancestral heroes of a particular and short narrative or descriptive lines, are recited at all important clan events, and single names from the praises may be used in greetings as well. The swallow is also frequently used in poetry to refer to someone who has gone far away and has subsequently returned.

The name **inhlolamvula** 'what predicts the rain' for martins overlaps with the name for swifts, as in Zulu folk taxonomy there is no difference between martins and swifts.

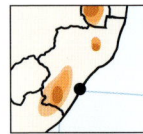

inkonjanesibhakabhaka Blue Swallow *Hirundo atrocaerulea*

The generic name **inkonjane** has been extended with *esibhakabhaka* 'sky-blue'.

inkonjanemnyama Black Saw-wing *Psalidoprocne holomelas*

The generic name **inkonjane** has been extended with *emnyama* 'which is black'.

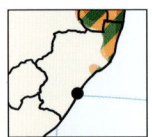

inkonjanemqolomlotha Grey-rumped Swallow *Pseudhirundo griseopyga*

The generic name **inkonjane** has been extended with *emqolomlotha* 'which is grey-rumped'.

inhlolamanzi Brown-throated Martin *Riparia paludicola*

The name **inhlolamanzi** means 'what predicts the water', a play on the generic name **inhlolamvula** 'what predicts the rain'.

umashwilomidwa Yellow-streaked Greenbul

igedezi Terrestrial Brownbul

inkonjanesibhakabhaka Blue Swallow

inkonjanemnyama Black Saw-wing

inkonjanemqolomlotha Grey-rumped Swallow

inhlolamanzi Brown-throated Martin

MARTIN & SWALLOW

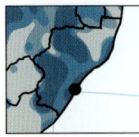

inhlolamfula Sand Martin *Riparia riparia*
The generic **inhlolamvula** has been changed by one letter (v → f) to produce the coinage **inhlolamfula** 'what examines the river', another play on words, and one that echoes the scientific name *Riparia riparia*, where 'riparia' refers to the river and its banks where these birds nest in colonies.

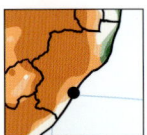

inhlolamvulebhandensundu Banded Martin *Riparia cincta*
The generic name **inhlolamvula** has been extended with *elibhandensundu* 'brown-banded'.

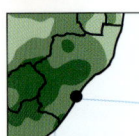

inhlolamvula yamadwala Rock Martin *Hirundo fuligula*
The generic name **inhlolamvula** has been extended with *yamadwala* 'of the rocks'.

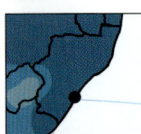

inkonjane yaseYurobhu Barn Swallow *Hirundo rustica*
The generic name **inkonjane** is extended here with *yaseYurobhu* 'from Europe', reflecting the bird's earlier name European Swallow.

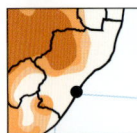

inkonjane yamawa South African Cliff Swallow *Pterochelidon spilodera*
The generic name **inkonjane** has been extended with *yamawa* 'of the cliffs'.

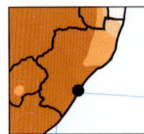

inkonjanemqalomhlophe White-throated Swallow *Hirundo albigularis*
The generic name **inkonjane** has been extended with *emqalomhlophe* 'white-throated'.

inhlolamvulelibhandensundu Banded Martin juv with beetle prey

inhlolamfula Sand Martin

inhlolamvulebhandensundu Banded Martin

inhlolamvula yamadwala Rock Martin

inkonjane yaseYurobhu Barn Swallow

inkonjane yamawa South African Cliff Swallow

inkonjanemqalomhlophe White-throated Swallow

SWALLOW & GRASSBIRD

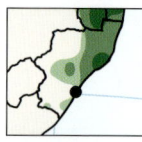

inkonjanesileside Wire-tailed Swallow *Hirundo smithii*
The generic name **inkonjane** has been extended with *esileside* 'long-tailed'.

inhlolamvula yasekhaya Common House Martin *Delichon urbacum*
The generic name **inhlolamvula** has been extended with *yasekhaya* 'of the home'.

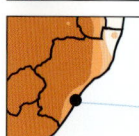

inkonjanenkulu Greater Striped Swallow *Cecropsis cucullata*
The generic name **inkonjane** has been extended with *enkulu* 'big'.

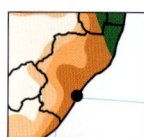

inkonjanencane Lesser Striped Swallow *Cecropsis abyssinica*
The generic name **inkonjane** has been extended with *encane* 'small'.

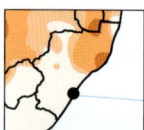

inkonjanesifubabomvu Red-breasted Swallow *Cecropsis semirufa*
The generic name **inkonjane** has been extended with *esifubabomvu* 'red-breasted'.

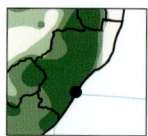

unontshiloza Cape Grassbird *Sphenoeacus afer*
The name **uontshiloza** is based on the verb *tshiloza* 'whistle, as a bird does', with the prefix –*no*–.

inkonjanesifubabomvu Red-breasted Swallow gathering mud to build nest

inkonjanesileside Wire-tailed Swallow

inhlolamvula yasekhaya Common House Martin

inkonjanenkulu Greater Striped Swallow

inkonjanencane Lesser Striped Swallow

inkonjanesifubabomvu Red-breasted Swallow

unontshiloza Cape Grassbird

CROMBEC & WARBLER

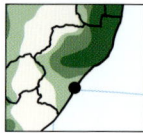

indibilishi Long-billed Crombec *Sylvietta rufescens*

The name **indibilishi** is long-established and widely-known. The name is a metaphor based on the noun *indibilishi* 'small coin', originally referring to a penny. The Zulu noun comes from the Dutch *dubbeltje*, a small coin worth two cents in Dutch currency. This small coin is probably a metaphor for the small size of this bird and its short stumpy tail (cf. Afrikaans *stompstert*).

This is not the only bird to be named after a coin. See the story of the 5c piece in the entry for the Blue Crane.

umqalaphuzi Yellow-throated Woodland Warbler *Phylloscopus ruficapilla*

This name is a compound of *umqalo* 'throat' and *ophuzi* 'which is yellow'.

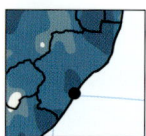

isicagogwane Willow Warbler *Phylloscopus trochilus*

The name **isicagogwane** has an earlier meaning of 'species of small grey bird which jumps from stick to stick' and this name is now assigned to the Willow Warbler.

REED WARBLER CLUSTER GENERIC NAME: ujamelumhlanga The name **ujamelumhlanga** means 'what stares at reeds'.

ujamelumhlangomncane Lesser Swamp Warbler *Acrocephalus gracilorostris*

The generic name **ujamelumhlanga** has been extended with *omncane* 'small'.

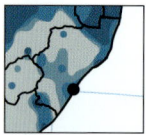

ujamelumhlangomkhulu Great Reed Warbler *Acrocephalus arundinaceus*

The generic name **ujamelumhlanga** has been extended with *omkhulu* 'which is big' for this bird.

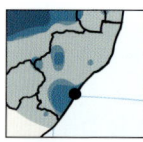

unomanduli Sedge Warbler *Acrocephalus schoenobaenus*

The name **unomanduli** uses both the –*no*– and –*ma*– prefixes before the noun *[i]nduli* which is a type of river grass or rush (*Cyperus* spp.) used in making mats. The name suggests a bird characteristically found among such rushes.

indibilishi Long-billed Crombec building a nest

indibilishi Long-billed Crombec

umqalaphuzi Yellow-throated Woodland Warbler

isicagogwane Willow Warbler

ujamelumhlangomncane Lesser Swamp Warbler

ujamelumhlangomkhulu Great Reed Warbler

unomanduli Sedge Warbler

WARBLER

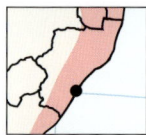

ujamelumhlanga waseYurobhu Eurasian Reed Warbler *Acrocephalus scirpaceus*

The generic name **ujamelumhlanga** is extended with *yaseYurobhu* 'from Europe'.

ujamelumhlanga wasemzansi African Reed Warbler *Acrocephalus baeticatus*

The generic name **ujamelumhlanga** is extended with *wasemzansi* 'of the lowlands'.

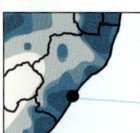

ujamelumhlanga wasenyakatho Marsh Warbler *Acrocephalus palustris*

The generic **ujamelumhlanga** is extended with *wasenyakatho* 'from the north'.

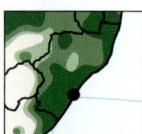

umavelashone Little Rush Warbler *Bradypterus baboecala*

The name **umavelashone** is a compound of the prefix *–ma–* 'characteristically' and the verbs *vela* 'appear' and *shona* 'disappear' and thus means 'the bird that is always appearing and disappearing'.

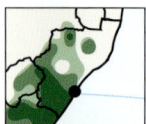

ujamelamafuku Barratt's Warbler *Bradypterus barratti*

The name **ujamelamafuku** plays cleverly on the generic name **ujamelumhlanga** by creating 'what stares at the thickets' (from *amafuku*, the plural of *ifuku* 'tangled thicket').

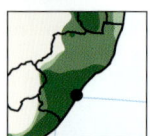

ujamelumhlangophuzi African Yellow Warbler *Iduna natalensis*

The generic name **ujamelumhlanga** is extended with *ophuzi* 'which is yellow'.

ujamelumhlanga wasemzansi African Reed Warbler nest

ujamelumhlanga waseYurobhu Eurasian Reed Warbler

ujamelumhlanga wasemzansi African Reed Warbler

ujamelumhlanga wasenyakatho Marsh Warbler

umavelashone Little Rush Warbler

ujamelamafuku Barratt's Warbler

ujamelumhlangophuzi African Yellow Warbler

WARBLER, GRASSBIRD, CISTICOLA & NEDDICKY

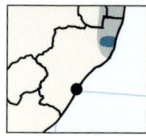

unomnqumo Olive-tree Warbler *Hippolais olivetorum*

The name **unomnqumo** is formed from the prefix –no– and *umnqumo*, a name for the wild olive *Olea europea africana*.

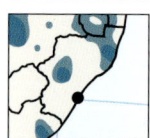

usikhothaphela Icterine Warbler *Hippolais icterina*

The name **usikhothaphela** is a compound of *khotha* 'catch' and *(i)phela* 'cockroach' and refers to the insectivorous diet of this bird.

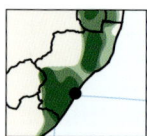

usotshanini Fan-tailed Grassbird *Schoenicola brevirostis*

The name **usotshanini** is formed from the prefix –so– and *[o]tshanini* (the locative form of *utshani* 'grass') and so means roughly 'the bird that is found in the grass'.

CISTICOLA GENERIC NAMES: ungcede, intinga The word '**ungcede**' has been found spelt in many ways (e.g. *ungcede, uncethe, unceda*) and has been translated as both 'lark' and 'warbler'. It is today the most commonly used word for the cisticola cluster of birds. It has no clear underlying meaning.

A Zulu proverb says *ungcede uthuma indlovu* (lit. 'the warbler sends the elephant'), i.e. you can get someone more powerful than you to do your bidding if you frame the request nicely. There is also a Zulu expression *amasi kangceda* ('the warbler's sour milk') for the solidified white humour found in the corner of the eye.

The name **intinga**, with no obvious underlying meaning, is also used commonly for cisticolas in general. In this book, both **uncede** and **intinga** are used to form species-specific names for cisticolas.

uncedobomvu Red-faced Cisticola *Cisticola erythrops*

The generic name **uncede** is extended with *obomvu* 'which is red'.

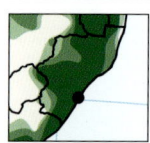

unovilane Lazy Cisticola *Cisticola aberrans*

This name is formed from the prefix –no– with the noun *ivila* 'lazy person' and the species-indicating suffix –ane and reflects 'lazy' in the English vernacular name.

uncede, ihlathinyane Neddicky *Cisticola fulvicapilla*

The generic name **uncede** is used specifically for this bird, as it is the historical source for the name 'Neddicky': Dutch-speaking farmers in the Eastern Cape first borrowed this word from the Xhosa language, and added the diminutive –tje to form 'ncedetje'. This in turn became the Afrikaans 'ncedetjie', taken over by English as 'neddicky'.

The name **ihlathinyane** is also used in some regions for this bird, a name formed from *ihlathi* 'forest' and the double diminutive –nyane, thus 'the very little bush species'.

unomnqumo Olive-tree Warbler

usikhothaphela Icterine Warbler

usotshanini Fan-tailed Grassbird

uncedobomvu Red-faced Cisticola

unovilane Lazy Cisticola

uncede Neddicky

CISTICOLA

iqobo Rattling Cisticola *Cisticola chiniana*
Iqobo is a well-known name for this bird, with no obvious underlying meaning.

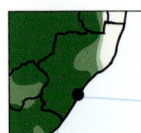

umdokwe Levaillant's Cisticola *Cisticola tinniens*
The name **umdokwe**, widely-used for this cisticola, appears to be unrelated to the noun *umdokwe* 'millet porridge', which is the base of the name **usamdokwe** for the Ring-necked Dove (see page 101).

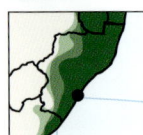

uncedoselesele Croaking Cisticola *Cisticola natalensis*
The generic name **uncede** is extended with *iselesele* 'frog', a reference to the croaking sound the bird makes, and thus echoing the English name.

intingelilayo Wailing Cisticola *Cisticola lais*
The generic name **intinga** is extended with *elilayo*, from the verb *lila* 'cry, weep, wail'.

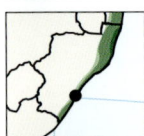

uncedomnyama Rufous-winged Cisticola *Cisticola galatotes*
The generic name **uncede** is extended with *omnyama* 'which is black' (cf. the Afrikaans *swartrugtinktinkie*).

uqhelu Zitting Cisticola *Cisticola juncidis*
The Zitting Cisticola is well-known as **uqhelu**, a name with no obvious underlying meaning.

uncedoselesele Croaking Cisticola ♀

iqobo Rattling Cisticola

umdokwe Levaillant's Cisticola

uncedoselesele Croaking Cisticola ♂

intingelilayo Wailing Cisticola

uncedomnyama Rufous-winged Cisticola

uqhelu Zitting Cisticola

CISTICOLA & PRINIA

uncede wehlane Desert Cisticola *Cisticola aridulus*
The generic name **uncede** is extended with *wehlane* 'of the desert'.

intingamafu Cloud Cisticola *Cisticola textrix*
The generic name **intinga** is extended with *amafu* 'clouds'.

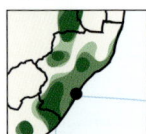

iqotshana Pale-crowned Cisticola *Cisticola cinnamomeus*
The name **iqotshana** is a diminutive form of the name **iqobo** assigned to the Rattling Cisticola (see previous page).

intinganqi Wing-snapping Cisticola *Cisticola ayresii*
The generic name **intinga** is extended with the ideophone *nqi* 'of snapping'.

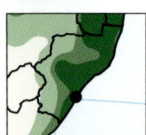

ujenga Tawny-flanked Prinia *Prinia subflava*
The name **ujenga** is taken from the Zulu word *ujenga* meaning 'species of small bird with long tail'.

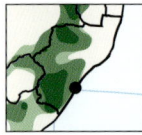

ujenga wokhahlamba Drakensberg Prinia *Prinia hypoxantha*
The name **ujenga** (see previous species) has been extended with *wokhahlamba* 'from the Drakensberg'.

iqotshana Pale-crowned Cisticola ♀

uncede wehlane Desert Cisticola

intingamafu Cloud Cisticola

iqotshana Pale-crowned Cisticola ♂

intinganqi Wing-snapping Cisticola

ujenga Tawny-flanked Prinia

ujenga wokhahlamba Drakensberg Prinia

APALIS, CAMAROPTERA, WREN-WARBLER & EREMOMELA

umabilwane Bar-throated Apalis *Apalis thoracica*
The name **umabilwane** is derived from the plural *amabilo* of the noun *ibilo*, with the suffix *-ane*. *Ibilo* refers to the dewlap of cattle and is also used for a double chin in humans. The bird's name is a reference to the marking on its throat which looks like a dewlap. The regional variation **umabhelwane** is also found for this bird.

umankole Rudd's Apalis *Apalis ruddi*
The name **umankole** is a variant, inspired by the call of the male bird, of the name **unkololo** – a well-known name (previously unrecorded) for Rudd's Apalis. Otherwise it has no apparent underlying meaning.

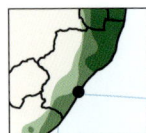
umankolophuzi Yellow-breasted Apalis *Apalis flavida*
The name **umankole** for Rudd's Apalis has been extended with *ophuzi* 'which is yellow'.

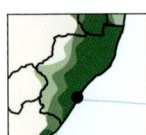
ibhoyi, imbuzane Green-backed Camaroptera *Camaroptera brachyura*
The name **ibhoyi** is a well-known and long-established name for this bird. It has no obvious underlying meaning.

The name **imbuzane**, also well-known and long-established is derived from *imbuzi* 'goat' and the suffix *–ane*, in reference to the bird's bleating alarm call.

isadube Stierling's Wren-Warbler *Calamonastes stierlingi*
This name compares the stripes on the bird to those of a zebra. The name is formed from *sa–* 'something like' and *idube* 'zebra'.

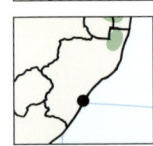
imbuzaneluhlaza Green-capped Eremomela *Eremomela scotops*
The name **imbuzane** has been extended with *eluhlaza* 'which is green'.

ibhoyi Green-backed Camaroptera nest

umabilwane Bar-throated Apalis

umankole Rudd's Apalis ♂

umankolophuzi Yellow-breasted Apalis

ibhoyi Green-backed Camaroptera

isadube Stierling's Wren-Warbler

imbuzaneluhlaza Green-capped Eremomela

EREMOMELA, BABBLER, BLACKCAP & WARBLER

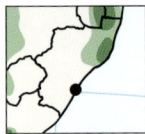
imbuzanephuzi Yellow-bellied Eremomela *Eremomela icteropygialis*
The name **imbuzane** has been extended with *ephuzi* 'which is yellow'.

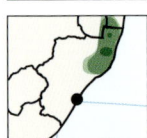
imbuzane yomnqawe Burnt-necked Eremomela *Eremomela usticollis*
The generic name **imbuzane** has been extended with *yomnqawe* 'of the acacia tree'.

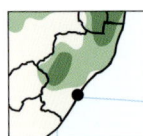
ihlekehle Arrow-marked Babbler *Turdoides jardineii*
The well-known and widely-used name **ihlekehle** is based on the two verbs *hleka* 'laugh' and *ehla* 'come down', 'alight'. The bird is seen as laughing (babbling) as it alights.

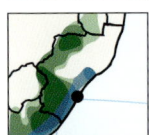
unosigqokomnyama Bush Blackcap *Lioptilus nigricapillus*
The name **unosigqokomnyama** echoes the English name, being a compound of –no– with *isigqoko* 'hat, cap' and *omnyama* 'which is black'.

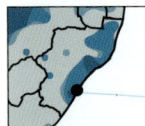
umnqumo Garden Warbler *Sylvia borin*
The word *umnqumo* refers to the wild olive *Olea europea africana*, a tree that frequently occurs in the bush where this bird is commonly found.

ihlekehleke Chestnut-vented Warbler *Sylvia subcaerulea*
The name **ihlekehleke** is based on a reduplication of the verb *hleka* 'laugh' with the same reference to song as in *ihlekehle* for the Arrow-marked Babbler above. The Afrikaans name *tjeriktik* also refers to its call.

ihlekehle Arrow-marked Babblers are sociable birds, juv have dull eyes

imbuzanephuzi Yellow-bellied Eremomela

imbuzane yomnqawe Burnt-necked Eremomela

ihlekehle Arrow-marked Babbler

unosigqokomnyama Bush Blackcap

umnqumo Garden Warbler

ihlekehleke Chestnut-vented Warbler

173

WHITE-EYE, MYNA & STARLING

WHITE-EYE GENERIC NAME: umehlwane and umbicini The white-eye cluster has two generic names. The name **umehlwane** is derived from *amehlo* 'eyes' and the suffix *–ane*, indicating both a species and a diminutive. The name thus means 'the little [bird] species with the [striking] eyes'.

The name **umbicini** is derived from *ubhici* 'purulent discharge from the eyes', and refers to the white ring around the eyes.

umehlwanoluhlaza, umbicini Cape White-eye *Zosterops virens*
The generic name **umehlwane** has been extended with *oluhlaza* 'which is green'. The name **umbicini** is also used for this bird.

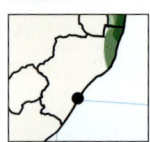

umehlwanophuzi, umbicini Southern Yellow White-eye *Zosterops anderssoni*
The generic name **umehlwane** has been extended with *ophuzi* 'yellow'. The name **umbicini** is also used.

usothathizwe Common Myna *Acridotheres tristis*
The name **usothathizwe** is based on *–so–* 'master of' with *thatha* 'take' and *izwe* 'country', i.e. the bird that is a master of taking over the whole country.

impofazana Wattled Starling *Creatophora cinerea*
A well-known and long-established name for this bird. It is derived from the adjective *mpofu* 'tawny', 'dun' with the suffixes *–azi* and *–ana*, together giving the meaning of 'tawnyish' and 'dunnish', in reference to the coloration of the bird.

ikhwinsi laseYurobhu Common Starling *Sturnus vulgaris*
The generic name **ikhwinsi** for the glossy starlings (see below) has been extended with *laseYorubhu* to create a name for this alien invasive species from Europe.

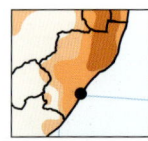

ikhwezelimacwebi Violet-backed Starling *Cinnyricinclus leucogaster*
The name **ikhwezi** (see below) has been extended with *(eli)macwebi*, probably derived from *ubucwebecwebe* 'sparkling, glittering splendour'.

ikhwezelimacwebi Violet-Backed Starling ♀

impofazana Wattled Starling ♀

umehlwanoluhlaza Cape White-eye

umehlwanophuzi Southern Yellow White-eye

usothathizwe Common Myna

impofazana Wattled Starling ♂

ikhwinsi laseYurobhu Common Starling

ikhwezelimacwebi Violet-backed Starling ♂

STARLING, OXPECKER & GROUND THRUSH

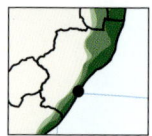

Ikhwezi lasogwini Black-bellied Starling *Notopholia corruscus*
The generic name **ikhwezi** has been extended with *lasogwini* 'from the coast'.

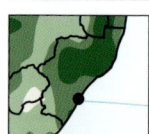

ikhwezi, ikhwinsi Cape Starling *Lamprotornis nitens*
Both these names are well-known and widely used in KwaZulu-Natal. The word *ikhwezi* also refers to the Morning Star which also glitters and shines. The name **ikhwinsi** is a variation.

ingwangwa, ikhwikhwi Pied Starling *Lamprotornis bicolor*
These two names are both well-known and widely used. Both are based on the reduplicated elements in the bird's call.

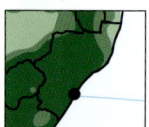

isomi Red-winged Starling *Onychognathus morio*
The name **isomi** is well-known and widely used and its variant form **insomi** is also used in various parts of KwaZulu-Natal. There is no special link (totemic) with the Msomi clan.

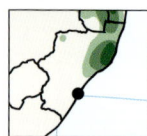

ihlalanyathi, ihlalankomo Red-billed Oxpecker *Buphagus erythrorhynchus*
The name **ihlalanyathi** means 'what sits on a buffalo' and the name **ihlalankomo** 'what sits on a head of cattle'. Both names are equally used today.

THRUSH GENERIC NAME: umunswi The name **umunswi**, with no obvious underlying meaning, is widely used for thrushes.

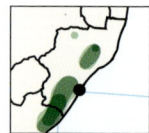

inswinswi Orange Ground Thrush *Geokichla gurneyi*
The name **inswinswi** is formed by repeating 'nswi' in 'umu-nswi'.

isomi Red-winged Starling ♂

ikhwezi lasogwini Black-bellied Starling

ikhwezi Cape Starling

ingwangwa Pied Starling

isomi Red-winged Starling ♀

ihlalanyathi Red-billed Oxpecker

inswinswi Orange Ground Thrush

GROUND THRUSH, THRUSH, ROBIN & ROBIN-CHAT

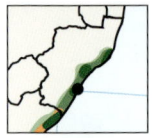

umunswi wehlathi Spotted Ground Thrush *Geokichla guttata*
The generic name **umunswi** has been extended with *wehlathi* 'of the forest'.

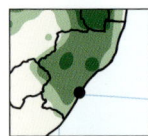

insansa Groundscraper Thrush *Turdus litsitsirupa*
An earlier meaning of this word was simply 'species of small bird, speckled black and white'. The word is assigned as a name for this thrush species.

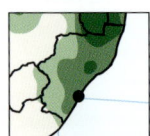

insansane Kurrichane Thrush *Turdus libonyana*
The name **insansa** (see Groundscraper Thrush above) has been extended with the suffix –ane.

umunswili Olive Thrush *Turdus olivaceus*
An extra syllable has been added to the generic name **umunswi** to form the name of this bird.

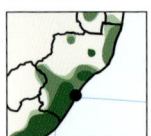

usonkanyezi White-starred Robin *Pogonocichla stellata*
The name for the bird is derived from *inkanyezi* 'star' with the prefix –so–, and so echoes both the specific epithet and the English vernacular.

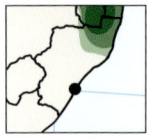

unomtshingo White-browed Robin-Chat *Cossypha heuglini*
This name is based on –no– and the noun *umtshingo* 'flute', in reference to the bird's well-known and characteristic flute-like call.

usonkanyezi White-starred Robin (juv)

umunswi wehlathi Spotted Ground Thrush

insansa Groundscraper Thrush

insansane Kurrichane Thrush

umunswili Olive Thrush

usonkanyezi White-starred Robin

unomtshingo White-browed Robin-Chat

ROBIN-CHAT & SCRUB ROBIN

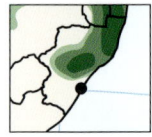

umbhekle White-throated Robin-Chat *Cossypha humeralis*

The name **umbhekle**, with no apparent underlying meaning, was earlier recorded for the Cape Robin-Chat, but is assigned to this bird.

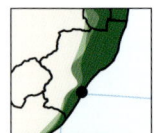

unonkositini Red-capped Robin-Chat *Cossypha natalensis*

The name **unonkositini** is formed from the prefix *–no–* and the noun *inkositini*, a word adopted from the English word 'concertina'. The name is a reference to the bird's musical song, which, with its incredible range of different sounds, resembles the versatility of a concertina.

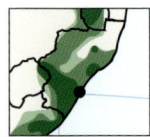

umananda Chorister Robin-Chat *Cossypha dichroa*

The name **umananda**, with no obvious underlying meaning, is commonly used for this bird.

ugaga Cape Robin-Chat *Cossypha caffra*

The name **ugaga** is well-known and widely-used for this bird. It has no obvious underlying meaning.

SCRUB ROBIN GENERIC NAME: ugagana The generic name **ugagana** is a diminutive form of the name **ugaga** (see above).

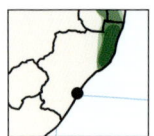

ugaganontshebe Bearded Scrub Robin *Cercotrichas quadrivirgata*

The generic name **ugagana** has been extended with *[i]ntshebe* 'beard' to form the name **ugaganontshebe**.

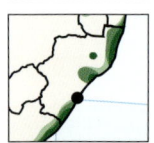

ugaganonsundu Brown Scrub Robin *Cercotrichas signata*

The generic name **ugagana** has been extended with *onsundu* 'which is brown' to form **ugaganonsundu.**

unonkositini Red-capped Robin-Chat (juv)

umbhekle White-throated Robin-Chat

unonkositini Red-capped Robin-Chat

umananda Chorister Robin-Chat

ugaga Cape Robin-Chat

ugaganontshebe Bearded Scrub Robin

ugaganonsundu Brown Scrub Robin

SCRUB ROBIN, STONECHAT, CHAT & WHEATEAR

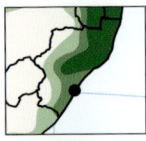

ugaganomidwa White-browed Scrub Robin *Cercotrichas leucophrys*
The generic name **ugagana** is extended with *omidwa* 'which is striped'.

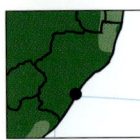

isichegu African Stonechat *Saxicola torquatus*
The name **isichegu**, with no underlying meaning, is well-known and widely used. It is sometimes found in its longer form **isichelegu**.

A historical name for this bird, no longer in common use, is **isangcaphela**, derived from the ideophone *ngcá* 'of brisk, spirited action' and *iphela* 'cockroach', referring to the way these bird snap up the insects commonly found on the ground.

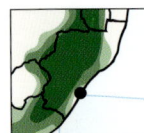

inkuletsheni Buff-streaked Chat *Campicoloides bifasciatus*
The name **inkuletsheni** is almost certainly derived from *inkweletsheni*, from *khwela* 'climb onto' and *etsheni* 'on the rock'. See **ikhwelentabeni** in the next entry.

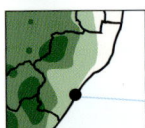

ikhwelentabeni Mountain Wheatear *Myrmecocihla monticola*
This name is based on *khwela* 'climb', 'climb onto' and *entabeni* 'on the mountain'.

umbexe Familiar Chat *Emarginata familiaris*
This name is well-known and widely-used for this bird. It has no apparent underlying meaning.

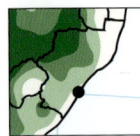

indlantuthwane Ant-eating Chat *Myrmecocichla formicivora*
This name follows the English vernacular name, the specific epithet (from Lat. *formex* 'ant' + *vora* 'eating'), and the generic name *Myrmecocichla* (from Gk. *myrmex* 'ant' + *cichla* 'thrush'). The Zulu name is derived from *dla* 'eat' and *intuthwane* 'ant'.

isichegu African Stonechat ♀

inkuletsheni Buff-streaked Chat ♀

ugaganomidwa White-browed Scrub Robin

isichegu African Stonechat ♂

inkuletsheni Buff-streaked Chat ♂

ikhwelentabeni Mountain Wheatear

umbexe Familiar Chat

indlantuthwane Ant-eating Chat

CLIFF CHAT, ROCK THRUSH & FLYCATCHER

isikhwelemaweni Mocking Cliff Chat *Thamnolaea cinnamomeiventris*
This name is derived from **khwela** 'climb onto' and *emaweni* 'on the cliffs'.

isihlalamatsheni Cape Rock Thrush *Monticola rupestris*
The name **isihlalamatsheni** means 'what lives (or sits) on the rocks'.

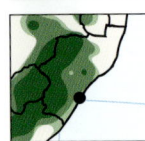

ikhwelematsheni Sentinel Rock Thrush *Monticola explorator*
The name **isikhwelematsheni** means 'what climbs onto the rocks'.

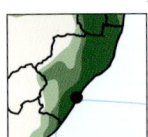

ummbesi Southern Black Flycatcher *Melaenornis pammelaina*
The name **ummbesi** is derived from the verb *embesa* 'clothe or cover someone with a blanket or cloak' suggesting that this dark-plumaged bird is covered in a 'cloak of darkness'.

isagundwane Pale Flycatcher *Melaenornis pallidus*
This name is derived from the prefix *–sa–* 'something like' and *igundwane* 'mouse', 'rat', and echoes the earlier English name Mouse-coloured Flycatcher.

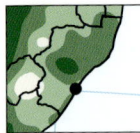

isaqola Fiscal Flycatcher *Melaenornis silens*
The name **isaqola** is formed from the prefix *–sa–* 'something like' and **iqola**, a name for the Southern Fiscal.

isikhwelemaweni Mocking Cliff Chat ♀

isihlalamatsheni Cape Rock Thrush ♀

isikhwelemaweni Mocking Cliff Chat ♂

isihlalamatsheni Cape Rock Thrush ♂

ikhwelematsheni Sentinel Rock Thrush ♂

ummbesi Southern Black Flycatcher

isagundwane Pale Flycatcher

isaqola Fiscal Flycatcher ♂

FLYCATCHER, TIT-FLYCATCHER, SUGARBIRD & SUNBIRD

usonambuzane Spotted Flycatcher *Muscicapa striata*

This name is derived from the prefix *–so–* 'father of', 'master of' with the noun *(isi) nambuzane* 'insect'. The name thus means 'the master of [catching] insect[s]'.

usikhothamlotha Ashy Flycatcher *Muscicapa caerulescens*

The noun *usikhotha* is derived from the verb *khotha* 'catch' (see also **inkotha** as a generic name for bee-eaters). It has been extended here with *–mlotha* 'ash-coloured'.

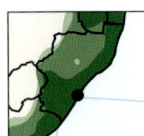

usikhothambuzane African Dusky Flycatcher *Muscicapa adusta*

This name is formed from *isikhotha* (see entry above) with *[isi]nambuzane* 'insect'.

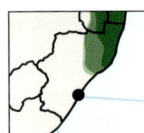

umantuluza Grey Tit-Flycatcher *Myioparus plumbeus*

This name, in all likelihood based on *–ma–* 'characteristically' and the verb *ntuluza* 'rush out rapidly', refers to feeding behaviour which is typical of flycatchers generally.

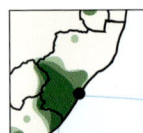

unosiqalaba Gurney's Sugarbird *Promerops gurneyi*

The name **unosiqalaba** is formed from the prefix *–no–* and the noun *isiqalaba* 'protea bush'.

SUNBIRD GENERIC NAME: incwincwi The names **incuncu** and **incwincwi** are regional forms of the same name. The name **incwincwi** is used as the generic name here, and both **incwincwi** and **incuncu** are used in extended form for some of the different species of sunbirds.

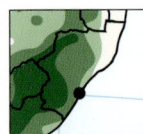

uhlazazana Malachite Sunbird *Nectarinia famosa*

The name **uhlazazana** is well-known and widely used for this bird. It is derived from *luhlaza* 'green' with the intensifying suffix *–azi* and the diminutive suffix *–ana*, thus 'the little intensely green bird'.

uhlazazana Malachite Sunbird ♀

usonambuzane Spotted Flycatcher

usikhothamlotha Ashy Flycatcher

usikhothambuzane African Dusky Flycatcher

umantuluza Grey Tit-Flycatcher

unosiqalaba Gurney's Sugarbird

uhlazazana Malachite Sunbird ♂

SUNBIRD

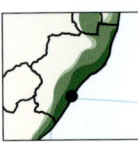

incwincweluhlaza Olive Sunbird *Cyanomitra olivacea*
The generic name **incwincwi** has been extended with *luhlaza* 'green'.

incwincwemphungana Grey Sunbird *Cyanomitra veroxii*
The generic **incwincwi** is extended with *mphunga* 'grey'. The diminutive suffix *–ana* is added to indicate a small bird.

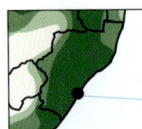

insusha Amethyst Sunbird *Chalcomitra amethystina*
The name of this bird is possibly derived from the word *insusha* meaning 'assegai with long shank and short blade', as this bird has a significantly long, straighter beak in comparison to several other sunbirds.

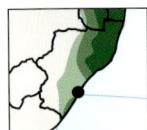

usifubabomvu Scarlet-chested Sunbird *Chalcomitra senegalensis*
The name **usifubabomvu** is formed from *isifuba* 'chest, breast' and *bomvu* 'red'.

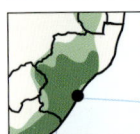

incuncwana Southern Double-collared Sunbird *Cinnyris chalybeus*
The generic name **incuncu** has been suffixed with the diminutive suffix *–ana*.

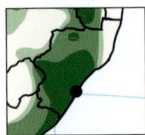

incuncu Greater Double-collared Sunbird *Cinnyris afra*
The generic name **incuncu** is used on its own for this bird.

usifubabomvu Scarlet-chested Sunbird ♀

incuncu Greater Double-collared Sunbird ♀

incwincweluhlaza Olive Sunbird

incwincwemphungana Grey Sunbird

insusha Amethyst Sunbird ♂

usifubabomvu Scarlet-chested Sunbird ♂

incuncwana Southern Double-collared Sunbird ♂

incuncu Greater Double-collared Sunbird ♂

SUNBIRD

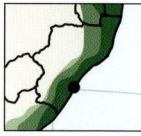

intonso, inqwathane Collared Sunbird *Hedydipna collaris*
The name **intonso** is probably derived from the verb *thonsa* meaning 'fall in drops', referring to the behaviour of this bird as it drops down after feeding on each flower. The underlying meaning of the other well-known name **inqwathane** is not apparent.

incwincwi yaseTembe Plain-backed Sunbird *Anthreptes reichenowi*
The generic name **incwincwi** has been extended with *yaseTembe* 'from Tembe' as in KwaZulu-Natal this bird is only found in the northern most parts, in the Tembe-Tonga area.

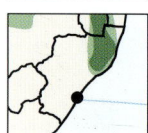

insonsi Marico Sunbird *Cinnyris mariquensis*
This name, with no apparent underlying meaning, is in oral usage for this bird.

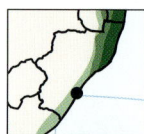

incwincwi yogu Purple-banded Sunbird *Cinnyris bifasciata*
The generic name **incwincwi** has been extended with *yogu* 'of the coast'.

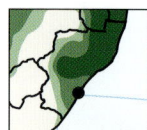

incwincwemhlophe White-bellied Sunbird *Cinnyris talatala*
The generic name **incwincwi** has been extended with *emhlophe* 'which is white'.

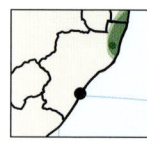

incwincwi yomfomothi Neergaard's Sunbird *Cinnyris neergaardi*
The generic name **incwincwi** has been extended with *yomfomothi* 'of the Lebombo Wattle' (*Newtonia hildebrandtii*), a tree with which this bird is associated.

incwincwi yogu Purple-banded Sunbird ♀

incwincwemhlophe White-bellied Sunbird ♀

intonso Collared Sunbird ♂

incwincwi yaseTembe Plain-backed Sunbird ♂

insonsi Marico Sunbird ♂

incwincwi yogu Purple-banded Sunbird ♂

incwincwemhlophe White-bellied Sunbird ♂

incwincwi yomfomothi Neergaard's Sunbird ♂

SPARROW & WEAVER

SPARROW GENERIC NAME: ujolwane The name **ujolwane** is derived from the word *ujolwana*, meaning a 'stay-at-home' type of person. It is used generically for sparrows.

ujolwane wekhaya House Sparrow *Passer domesticus*
The generic name **ujolwane** has been extended with *wekhaya* 'of home'.

undlunkulu Cape Sparrow *Passer melanurus*
The name **undlunkulu** is a well-known and widely-used name for this bird. Derived from *indlu* 'hut, house' and *khulu* 'big', the name refers to the exceptionally large, untidy nest built by this bird.

ujolwanokhandaphunga Southern Grey-headed Sparrow *Passer diffusus*
The generic name **ujolwane** has been extended with *ekhandaphunga* 'which is grey-headed'.

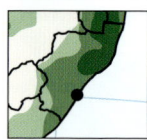

ujolwane wehlanze Yellow-throated Petronia *Gymnoris superciliaris*
The generic name **ujolwane** has been extended with *wehlanze* 'of the bushveld'.

WEAVER GENERIC NAME: ihlokohloko The name **ihlokohloko** is well-known and widely used as a generic name for the yellow weavers. The name is onomatopoeic referring to the various chattering, wheezing and bubbling call of these birds. Note that the expression '*Kanti kwakhele amahlokohloko lapha endlini na*?' (Literally: Is there a lot of weaver-bird nest-building going on in the hut?) is used when reproaching a lot of noisy children. This expression was recorded over a hundred years ago and is still in use today. The name **igelekeshe**, which is probably also onomatopoeic, is used as well throughout KwaZulu-Natal as a generic name for the yellow weavers.

usiqhophokezi Thick-billed Weaver *Amblyospiza albifrons*
The name **isiqhophokezi** is said to be derived from a noun meaning a protruding enlargement, like an outcrop on a hill, or a gall on a tree, and is a reference to the thick, hard bill.

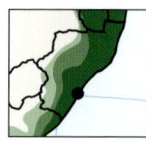

uzibukwana Spectacled Weaver *Ploceus ocularis*
The name **uzibukwana** is derived from a diminutive form of *izibuko* 'spectacles' and so means 'the little one with glasses'. The Zulu name thus echoes the English, Afrikaans (*brilwewer*) and Latin (*ocularis*) names.

ujolwane wekhaya House Sparrow ♂

undlunkulu Cape Sparrow ♂

ujolwanokhandaphunga Southern Grey-headed Sparrow

ujolwane wehlanze Yellow-throated Petronia

usiqhophokezi Thick-billed Weaver ♂

uzibukwana Spectacled Weaver ♂

193

WEAVER

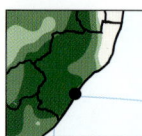
ihlokohlokelikhulu Cape Weaver *Ploceus capensis*
The generic name **ihlokohloko** has been extended with *elikhulu* 'which is big'.

GOLDEN WEAVER GENERIC NAME: igelesha The name **igelesha** is a shortened form of the name **igelekeshe** mentioned above as a known word for yellow weavers throughout KwaZulu-Natal.

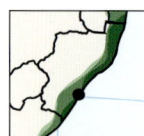
igelesha logu Eastern Golden Weaver *Ploceus subaureus*
The generic name **igelesha** has been extended with *logu* 'of the coast'.

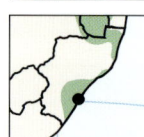
igelesha lehlathi Holub's Golden Weaver *Ploceus xanthops*
The generic name **igelesha** has been extended with *lehlathi* 'of the forest'.

igeleshelimqalonsundu Southern Brown-throated Weaver *Ploceus xanthopterus*
The generic name **igelesha** has been extended with *elimqalonsundu* 'brown-throated'.

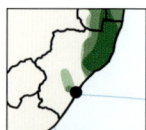
ihlokohlokwana Lesser Masked Weaver *Ploceus intermedius*
The generic name **ihlokohloko** has been extended with the diminutive suffix *–ana*.

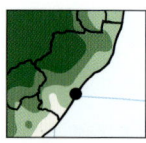
umzwingili Southern Masked Weaver *Ploceus velatus*
The name **umzwingili** has no apparent underlying meaning. It may be onomatopoeic.

ihlokohlokwana Lesser Masked Weaver ♀

umzwingili Southern Masked Weaver ♀

ihlokohlokelikhulu Cape Weaver ♂

igelesha logu Eastern Golden Weaver ♂

igelesha lehlathi Holub's Golden Weaver ♂

igeleshelimqalonsundu Southern Brown-throated Weaver ♂

ihlokohlokwana Lesser Masked Weaver ♂

umzwingili Southern Masked Weaver ♂

WEAVER & QUELEA

ihlokohloko lomuzi Village Weaver *Ploceus cucullatus*
The generic name **ihlokohloko** has been extended with *lomuzi* 'of the homestead'.

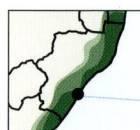

ihlokohloko lehlathi, idonsi Dark-backed Weaver *Ploceus bicolor*
The generic name **ihlokohloko** is extended with *lehlathi* 'of the forest'. The name **idonsi**, with no apparent underlying meaning, is also a well-known and widely-used name.

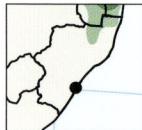

ukhandaklebhu Red-headed Weaver *Anaplectes rubriceps*
This name is a compound of the noun *ikhanda* 'head' and the ideophone *klébhu* 'of bright red colour'.

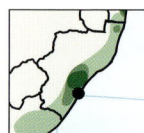

ukhandabomvu Red-headed Quelea *Quelea erythrops*
The name **ukhandabomvu** is a compound of *ikhanda* 'head' and *bomvu* 'red'.

isicibilili Red-billed Quelea *Quelea quelea*
The name **isicibilili** has the underlying meaning of 'species of small brown bird with red beak' and is applied here to the Red-billed Quelea.

ihlokohloko lomuzi Village Weaver ♀

ukhandabomvu Red-headed Quelea ♀

ihlokohloko lomuzi Village Weaver ♂

ihlokohloko lehlathi Dark-backed Weaver ♂

ukhandaklebhu Red-headed Weaver ♂

ukhandabomvu Red-headed Quelea ♂

isicibilili Red-billed Quelea ♂

isicibilili Red-billed Quelea ♀

197

BISHOP & WIDOWBIRD

BISHOP, WIDOWBIRD, WHYDAH GENERIC NAMES The name **intaka** is a generic for a large group of finch-like birds. When these have species-specific names, this is often for the male only, with the female being referred to as **intaka**. In the entries that follow both male and female names will be given where these are different and will be marked with the gender signs ♂ and ♀. It can be noted that **intaka** is also the Xhosa word for 'bird'.

BISHOP GENERIC NAME: isigwe The name **isigwe** is a long-established name for the Southern Red Bishop, and is used here generically for all bishop birds.

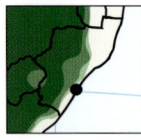

intakanyosi Yellow-crowned Bishop *Euplectes afer*

This name is a compound of **intaka** 'finch' and *inyosi* 'bee', a reference to the bumble-bee-like appearance when the bird is puffed up when males are displaying.

In some parts of KwaZulu-Natal the name **untumane** is also used, a name derived from *intuma*, the berry of the bitter apple shrub *Solanum incanum*. This fruit has a similar colouring as the bird.

ibomvana, isigwe, intakansinsi Southern Red Bishop *Euplectes orix*

All three of these names are well-known and widely-used. The name **ibomvana** is from the adjective *bomvu* ('red') and the diminutive suffix –*ana* and thus means 'the little red one'. The name **isigwe** has been selected as the generic name for bishops, and is used here unextended. The name **intakansinsi** is a compound of *intaka* 'finch' and *insinsi* 'seed of the *Erythrina* tree *umsinsi*'. The bird has the same black and red colouring as the seed.

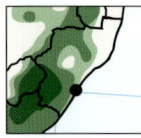

umambathilanga Yellow Bishop *Euplectes capensis*

This name is a compound of the verb *mbatha* 'wear a blanket over the shoulder' and *ilanga* 'sun'. It is a reference to the bright yellow on the bird's shoulders and back.

INDIGOBIRD AND WIDOW GENERIC NAME: umfelokazi The noun **umfelokazi** is the Zulu word for 'widow' so its meaning has been broadened here to include the birds apparently dressed in 'widow's weeds'. The base of the name is the verb *fa* ('die'), from which the noun *umfelo* ('the bereaved one') is derived. To this is added the feminine suffix –*kazi*.

umangube ♂, intaka ♀ Fan-tailed Widowbird *Euplectes axillaris*

The names **umangube** and **umahube**, with no apparent underlying meaning, are regional variants of the same word, with **umangube** being the more widely used.

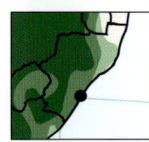

ibhaku ♂, isakabuli ♂, intaka ♀ Long-tailed Widowbird *Euplectes progne*

The names **ibhaku** and **isakabuli** are well-known and long-established names for the male of this species. **Ibhaku** comes from the ideophone *bháku* 'of flapping, fluttering wings'. The name **isakabuli** has no apparent underlying meaning.

As with many of the birds in the widowbird cluster, the female is called **intaka**.

ujojo ♂, intaka ♀ Red-collared Widowbird *Euplectes ardens*

Both names are well-known and widely-used names for the male and female of this species.

There is a Zulu proverb '*Ujojo umi ngothi lwakhe*' (literally: the jojo finch stands by means of his own stick) in other words people tend to stick with their own beliefs and decisions.

intakanyosi Yellow-crowned Bishop ♂

ibomvana Southern Red Bishop ♂

umambathilanga Yellow Bishop ♂

umangube Fan-tailed Widowbird ♂

ibhaku Long-tailed Widowbird-♂

ujojo Red-collared Widowbird-♂

199

WIDOWBIRD, PYTILIA, TWINSPOT & FIREFINCH

intakemaphikamhlophe White-winged Widowbird *Euplectes albonotatus*

In this name the generic **intaka** has been extended with *emaphikamhlophe* 'which is white-winged', from *amaphiko* 'wings' and *mhlophe* 'white'.

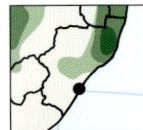

usantiyane Green-winged Pytilia *Pytilia melba*

This name is formed from the prefix *–sa–* 'something like' and **intiyane**, the name of the Common Waxbill.

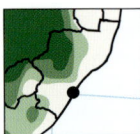

ugazini Red-headed Finch *Amadina erythrocephala*

This name, literally meaning 'in blood', refers to the red markings on the bird. It is derived from *igazi* 'blood'.

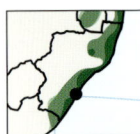

intiyaneluhlaza Green Twinspot *Mandingoa nitidula*

In this name, **intiyane** (the name for the Common Waxbill) is extended with *luhlaza* 'green'.

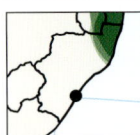

umagumejana Pink-throated Twinspot *Hypargos margaritatus*

This coined name is derived from the prefix *–ma–* with the clan name Gumede and the diminutive suffix *–ana*. The reference is to a young Gumede girl renowned for her decorative beadwork. The pearl-like spots on this bird are thus obliquely referred to in the names from three languages, with the Zulu bead-work reference, the English 'twinspot' referring to the two spots on individual feathers of this bird, the genus name *Hypargos* ('a hundred eyes below') which refers to Greek goddess Hera putting the hundred eyes of Argos onto the tail of the peacock, and the specific epithet *margaritatus* which is derived from the Latin word for a pearl.

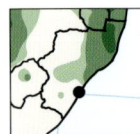

inkashana Red-billed Firefinch *Lagonosticta senegala*

This known and currently used name is clearly an abbreviation of the earlier-recorded name **inkayishana**, with no apparent underlying meaning.

This was also the name of uNkayishana Solomon kaDinuzulu, the king of the Zulus from 1913 to 1933 and grandfather of the current king Goodwill Zwelithini kaBhekuzulu.

intakemaphikamhlophe White-winged Widowbird ♀

usantiyane Green-winged Pytilia ♀

intakemaphikamhlophe White-winged Widowbird ♂

usantiyane Green-winged Pytilia ♂

ugazini Red-headed Finch ♂

intiyaneluhlaza Green Twinspot ♂

umagumejana Pink-throated Twinspot ♂

inkashana Red-billed Firefinch ♂

FIREFINCH & WAXBILL

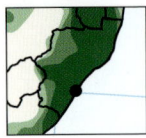

ubucubu African Firefinch *Lagonosticta rubricata*

The name **ubucubu** is well-known and widely-used for this bird.

This bird name is also 'jocularly' applied to children below the age of about five years. To illustrate this, there is an expression *kangafumanisa 'muntu, ubucubu bodwa*, meaning 'I didn't find a single person at home, only little children'. There is also the following admonition to a small child: *ub'uhamba wedwa nje; kawazi yini ukuthi ubucubu buhamba ngababili na?*, meaning 'you were just going alone; don't you know that waxbills (i.e. little children) always go in pairs?'

insewane Jameson's Firefinch *Lagonosticta rhodopareia*

The name **insewane** is likely to be another regional variant of **intiyane** or **insiyane**, used generically for waxbills.

WAXBILL GENERIC NAME: intiyane The name **intiyane**, with no apparent underlying meaning, is a long-established and widely-used generic name for the waxbills. In some regions the form **insiyane** is used.

isicelankobe Blue Waxbill *Uraeginthus angolensis*

The name **isicelankobe** is formed from the verb *cela* 'ask for, request' and *[izi]nkobe* 'sorghum or mealie grains, especially when boiled'. This is a bird frequently found in and around Zulu homesteads, attracted to the small grain particles spilt when sorghum or maize is being ground.

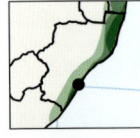

ivuzigazi Grey Waxbill *Estrilda perreini*

The name **ivuzigazi**, which means 'what leaks blood', describes perfectly the red colour of the rump of the bird.

intiyanejwayelekile Common Waxbill *Estrilda astrild*

The generic form **intiyane** is extended here with *ejwayelekile* 'common, well-known'.

Another name is **indlovuyenduna** 'bull elephant'. The cry of this tiny bird is said to be *Ngingangendlovu yenduna* 'I am as big as a bull elephant!' There is a certain humour in this name, similar to the name **ingangomfula** 'as big as the river', a name for the Natal Kingfisher (see entry for the Pygmy Kingfisher above), but there may be a good reason for this name as well: It is possible that this name may refer to the habit of these birds when disturbed of flying up to approximately the height of a large elephant, before flying down to the ground again. The height of an elephant is also the level at which they customarily fly.

ubusukuswane Swee Waxbill *Coccopygia melanotis*

The name **ubusukuswane**, with no apparent underlying meaning, is a widely-used and long-established name for this bird.

ubucubu African Firefinch ♂

insewane Jameson's Firefinch ♂

isicelankobe Blue Waxbill ♂

ivuzigazi Grey Waxbill

intiyanejwayelekile Common Waxbill

ubusukuswane Swee Waxbill ♂

Ingrid Weiersbye

WAXBILL, QUAIL FINCH, MANNIKIN & INDIGOBIRD

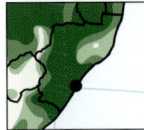

isabhonsi Orange-breasted Waxbill *Amandava subflava*

The name **isabhonsi** is derived from the prefix –sa– 'something like' and *ibhonsi*, which is the fruit of the tree *Salacia kraussii*. The colour of the bird's breast is very like that of the ripe fruit.

inxenge, unonklwe Quailfinch *Ortygospiza atricollis*

The name **inxenge** has no apparent underlying meaning. The other name **unonklwe** appears to be onomatopoeic, and is in current usage in some areas.

MANNIKIN GENERIC NAME: amadojeyana This name is a double diminutive of the word **amadoda** 'men', and given in the plural, as these birds are always in little flocks. The name means 'a number of tiny men'.

amadojeyanajwayelekile Bronze Mannikin *Lonchura cucullata*

The generic name **amadojeyana** has been extended with *ajwayelekile* 'common', 'well-known'.

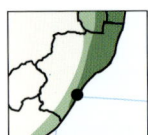

amadojeyanabomvu Red-backed Mannikin *Lonchura nigriceps*

The generic name **amadojeyana** has been extended with *bomvu* 'red'.

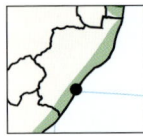

amadojeyanalunga Magpie Mannikin *Lonchura fringilloides*

The generic name **amadojeyana** has been extended with *(i)lunga* 'black-and-white beast'. This word *ilunga*, also the name of the Southern Fiscal, is used in Zulu bird names in the same way as the English word 'pied' is used.

umfelokazomlomobomvu Village Indigobird *Vidua chalybeata*

The generic **umfelokazi** for widowbirds has been extended with *omlomobomvu* 'which is red-mouthed'.

amadojeyanajwayelekile Bronze Mannikin (juv)

umfelokazomlomobomvu Village Indigobird ♀

isabhonsi Orange-breasted Waxbill ♂

inxenge Quailfinch ♂

amadojeyanajwayelekile Bronze Mannikin

amadojeyanabomvu Red-backed Mannikin

amadojeyanalunga Magpie Mannikin

umfelokazomlomobomvu Village Indigobird ♂

CUCKOO FINCH, INDIGOBIRD, WHYDAH & WAGTAIL

unondindwa Cuckoo Finch *Anomalospiza imberbis*

For the name of this bird, the word *unondindwa* has been used. It means 'a girl out of parental control; a woman who leads an immoral town life', inferring a woman who has babies but leaves them to others to bring up, as indeed does this parasitic finch.

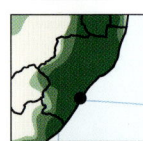

umfelokazomlomomhlope Dusky Indigobird *Vidua funerea*

The generic name **umfelokazi** has been extended with *omlomomhlophe* 'which is white-mouthed'.

uhlekwane, ojojekhaya Pin-tailed Whydah *Vidua macroura*

The name **uhlekwane** is conceivably derived from the noun *uhleko* and the suffix *–ane*. *Uhleko* is itself derived from the verb *hleka* laugh, and in this usage *uhleko* refers to high-pitched laughter of young girls. The bird's call is similar.

The name **ujojo** is applied to the Red-collared Widowbird (see above) and it is here extended with *ekhaya* 'at home' to form a name well-known for the Pin-tailed Whydah, a bird frequently found around homes.

ibhaku lehlanze Long-tailed Paradise Whydah *Vidua paradisaea*

The name **ibhaku** is recorded for the Long-tailed Widowbird *Euplectes progne* and is extended with *lehlanze* 'of the bushveld' for this similarly long-tailed bird.

WAGTAIL GENERIC NAME: umvemve The name **umvemve**, with no apparent underlying meaning, is a long-established and widely-used generic name for wagtails. In some regions the name **umncishu**, with no obvious underlying meaning, is also used.

In both Xhosa and Zulu traditional culture, wagtails are revered and seen as lucky birds that ensure that a man's herd of cattle prosper and grow. To harm or kill a wagtail is regarded as a sacrilege and young boys are warned about this.

In the Zulu folktale *Inyoni Yamasi* 'the 'maasbird' it is the wagtail that magically fills up the calabashes of a poor family with sour milk whenever they want.

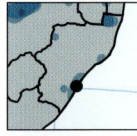

umvemvophuzi Western Yellow Wagtail *Motacilla flava*

The generic name **umvemve** has been extended with *ophuzi* 'which is yellow'.

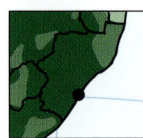

umncishu Cape Wagtail *Motacilla capensis*

The name **umncishu** is used for wagtails in various parts of KwaZulu-Natal and is assigned here specifically to the Cape Wagtail.

unondindwa Cuckoo Finch ♂

umfelokazomlomomhlope Dusky Indigobird ♂

uhlekwane Pin-tailed Whydah ♂

ibhaku lehlanze Long-tailed Paradise Whydah ♂

umvemvophuzi Western Yellow Wagtail

umncishu Cape Wagtail

WAGTAIL, LONGCLAW & PIPIT

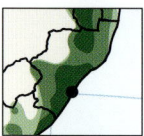

umvemventaba Mountain Wagtail *Motacilla clara*

The generic name **umvemve** has been extended with *[i]ntaba* 'mountain'.

umvemvolunga African Pied Wagtail *Motacilla aguimp*

The generic name **umvemve** has been extended with *olunga*. This is another example of the word *ilunga*, meaning both 'a black-and-white beast' and Southern Fiscal, to indicate pied plumage.

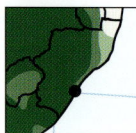

inqomfi Cape Longclaw *Macronyx capensis*

The name **inqomfi**, with no apparent underlying meaning, is a well-known and widely-used name for this bird. This bird occurs as a subject in both traditional oral poetry as well as modern Zulu written poetry. In oral praises for this bird, collected by James Stuart more than a hundred years ago, a deeply allusive line suggests that the plumage of this bird has been dyed red by dye obtained from the roots of the *intolwane* plant (*Elephantorhiza elephantina*). A well-known phrase in Zulu – *uJamludi obomvu njengentolwane*, meaning 'Jamludi the blood-coloured ox who is red like the *intolwane*' – emphasizes the link between this plant and the '*bomvu*' colour.

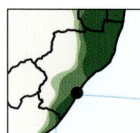

itoyiya Yellow-throated Longclaw *Macronyx croceus*

This bird has accumulated a number of Zulu names over the years, including **inqomfi**, **igwili**, **igwilintsi**, **unoyihomboyi**, **unotoyi** and **itoyiya**. Of these, the name **itoyiya**, with no clear underlying meaning, is used for this bird.

The name **igwili**, which refers to the yellow colour, is still used, but not commonly. It is mainly restricted to the Vryheid/Newcastle area. The name **igwilintsi** refers to the yellow colour and the call, and is still used in the areas where the name **igwili** is found.

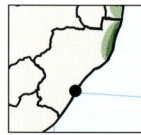

inqomfebomvu, unotoyobomvu Rosy-throated Longclaw *Macronyx ameliae*

Two different names are used for this bird. The name **inqomfebomvu** is the generic name **inqomfi** with *bomvu* 'red'. **Unotoyobomvu** is the name **unotoyi**, one of several names recorded for wagtails, also with *bomvu* 'red'.

PIPIT GENERIC NAME: umngcelu The names **umngcelu** and **ingcelekeshe** (a longer variant of **umngcelu**) are long-established names for pipits. Neither has any apparent underlying meaning.

A well-known Zulu expression is '*ngizovuka imingcelu ingakaqali*' (lit. 'I shall wake, the pipits not yet having started'), i.e. very early in the morning, as these birds are supposed to be amongst the first to start singing at the break of day.

ingcelekeshe African Pipit *Anthus cinnamomeus*

The name **ingcelekeshe** is almost certainly derived from the Zulu ideophone *gcélekeshe* 'of coming suddenly into the open', 'of darting out'.

umvemventaba Mountain Wagtail

umvemvolunga African Pied Wagtail

inqomfi Cape Longclaw

itoyiya Yellow-throated Longclaw

inqomfebomvu Rosy-throated Longclaw

ingcelekeshe African Pipit

PIPIT

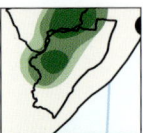

umngcelu wentaba Mountain Pipit *Anthus hoeschi*
The generic name **umngcelu** has been extended with *wentaba* 'of the mountain'.

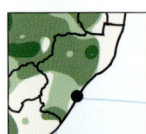

umngcelu wehlungu Buffy Pipit *Anthus vaalensis*
The generic name **umngcelu** has been extended with *wehlungu* 'of the newly burnt veld'.

umngcelu wegquma Nicholson's Pipit *Anthus nicholsoni*
The generic name **umngcelu** has been extended with *wegquma* 'of the hillock', a reference to the lower-lying foothills where this bird is mostly found.

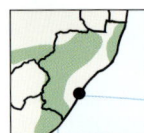

umngcelwana Short-tailed Pipit *Anthus brachyurus*
The generic name **umngcelu** has been extended with the diminutive suffix –*ana*.

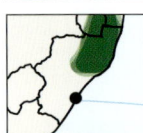

umngcelohlanze Bushveld Pipit *Anthus caffer*
The generic name **umngcelu** has been extended with *ohlanze* 'of the bushveld'.

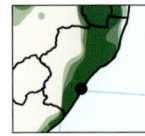

umngcelomidwa Striped Pipit *Anthus lineiventris*
The generic name **umngcelu** has been extended with *omidwa* 'which is striped'.

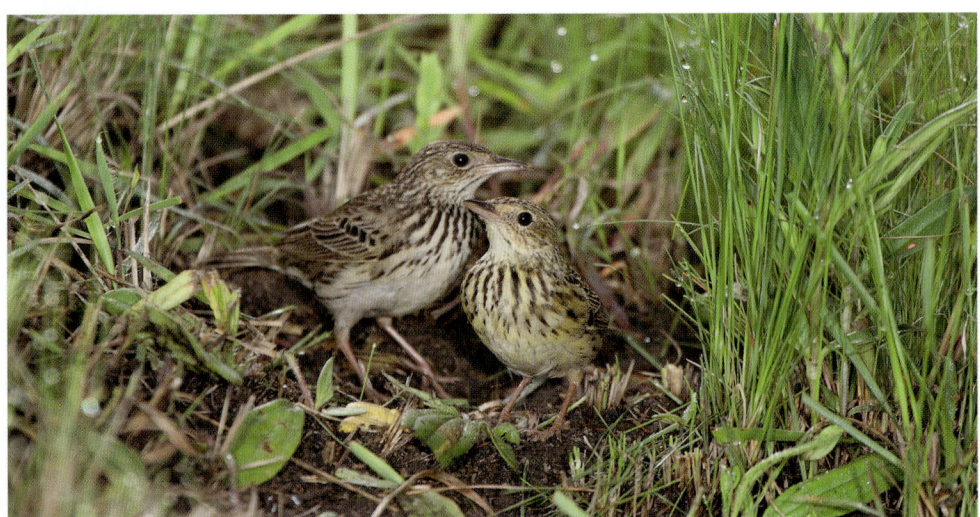

umngcelwana Short-tailed Pipit ♂ and ♀

umngcelu wentaba Mountain Pipit

umngcelu wehlungu Buffy Pipit

umngcelu wegquma Nicholson's Pipit

umngcelwana Short-tailed Pipit ♂

umngcelohlanze Bushveld Pipit

umngcelomidwa Striped Pipit

Adam Riley - Rockjumper Birding

PIPIT, SEEDEATER & CANARY

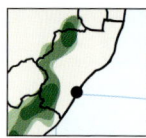

ungcelekeshephuzi Yellow-breasted Pipit *Anthus chloris*
The name **ingcelekeshe** has been extended with *ephuzi* 'which is yellow'.

Recent research proposes that this species is better placed in the genus *Macronyx* Longclaws.

umngceloze Plain-backed Pipit *Anthus leucophrys*
The generic name **umngcelu** has been suffixed with *–ze*, signifying 'with nothing', a reference to the lack of distinguishing features displayed by this bird. (See 'plain' in the English name.)

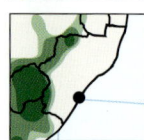

umncgelu wasematsheni African Rock Pipit   *Anthus crenatus*
The generic name **umngcelu** has been extended with **wasematsheni** 'from among the rocks'.

CANARY GENERIC NAME: umbhalane The name **umbhalane** is well-known for birds in the canary cluster. It is most likely derived from the verb *bhala* 'write', 'make a mark', a reference to the yellow or white stripe above the eye (supercilium).

There is a Zulu saying '*Uyoze ube nobala njengombhalane*' (literally: 'You will eventually have a mark like the wild canary'), i.e. You will be a marked person [if you do such and such].

umbhalanonsundu Streaky-headed Seedeater *Crithagra gularis*
The generic name **umbhalane** has been extended with *onsundu* 'which is brown'.

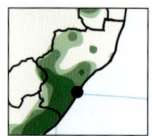

umbhalane wehlathi Forest Canary *Crithagra scotops*
The generic name **umbhalane** has been extended with *wehlathi* 'of the forest'.

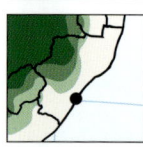

unogilomnyama Black-throated Canary *Crithagra atrogularis*
The name **unongilomnyama** is formed from the prefix *–no–* with *ingilo* 'throat' and *mnyama* 'black'.

umbhalane wehlathi Forest Canary ♂ drinking

ungcelekeshephuzi Yellow-breasted Pipit ♂

umngceloze Plain-backed Pipit

umncgelu wasematsheni African Rock Pipit

umbhalanonsundu Streaky-headed Seedeater

umbhalane wehlathi Forest Canary ♂

unogilomnyama Black-throated Canary

CANARY & SISKIN

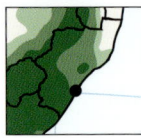

umzwilili Cape Canary *Serinus canicollis*
This onomatopoeic name is well-known and widely used for this bird. In some regions the variant form **umzwili** is also used.

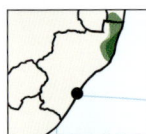

umalaleni Lemon-breasted Canary *Crithagra citrinispectus*
This name is derived from *emalaleni*, the locative form of *amalala* 'lala palms'. This bird is commonly found in Palm Veld, and nests in the Lala Palm (*Hyphaene coriacea*).

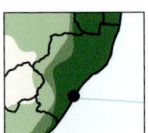

umbhalanomadevu Yellow-fronted Canary *Crithagra mozambica*
The generic name **umbhalane** is extended with *omadevu* 'whiskered', 'having a moustache'.

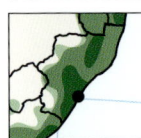

umbhalanomkhulu Brimstone Canary *Crithagra sulphurata*
The generic name **umbhalane** has been extended with *omkhulu* 'which is big'.

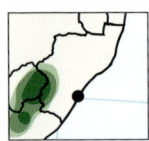

umbhalane wokhahlamba Drakensberg Siskin *Crithagra symonsi*
The generic name **umbhalane** has been extended with *wokhahlamba* 'of the Drakensberg'.

umbhalane wokhahlamba Drakensberg Siskin ♀

BUNTING

BUNTING GENERIC NAME: usokhandamidwa The generic name **usokhandamidwa** is based on the prefix –so–, the noun *ikhanda* 'head' and *[i]midwa* 'stripes'. The name means 'the bird that is particularly striped on the head'.

usokhandamidwa wamatshe Cinnamon-breasted Bunting *Emberiza tahapisi*

The generic name **usokhandamidwa** is extended with *wamatshe* 'of the rocks'.

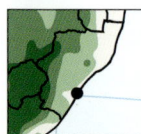

usokhandamidwonsundu Cape Bunting *Emberiza capensis*

The generic name **usokhandamidwa** has been extended with *onsundu* 'which is brown'.

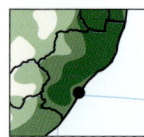

usokhandamidwombalabala Golden-breasted Bunting *Emberiza flaviventris*

The generic name **usokhandamidwa** has been extended with *ombalambala* 'which is multi-coloured'.

usokhandamidwombalabala Golden-breasted Bunting (juv)

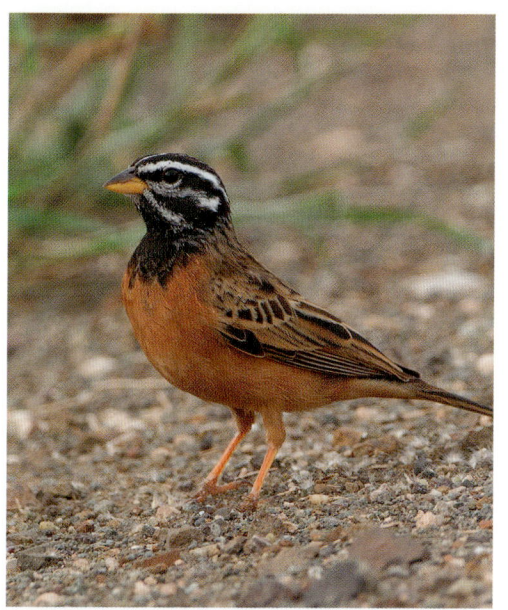

usokhandamidwa wamatshe Cinnamon-breasted Bunting

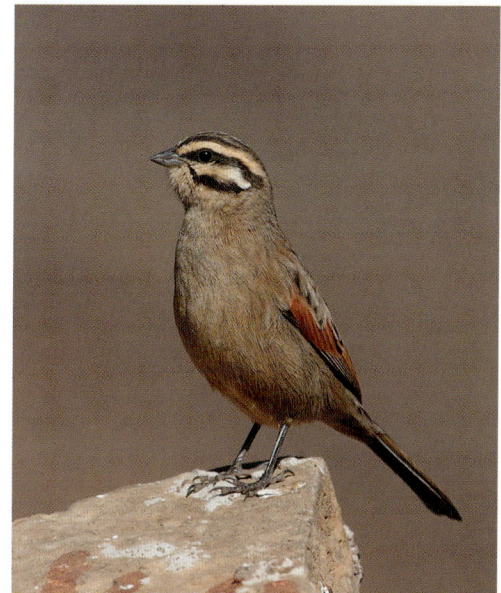

usokhandamidwonsundu Cape Bunting

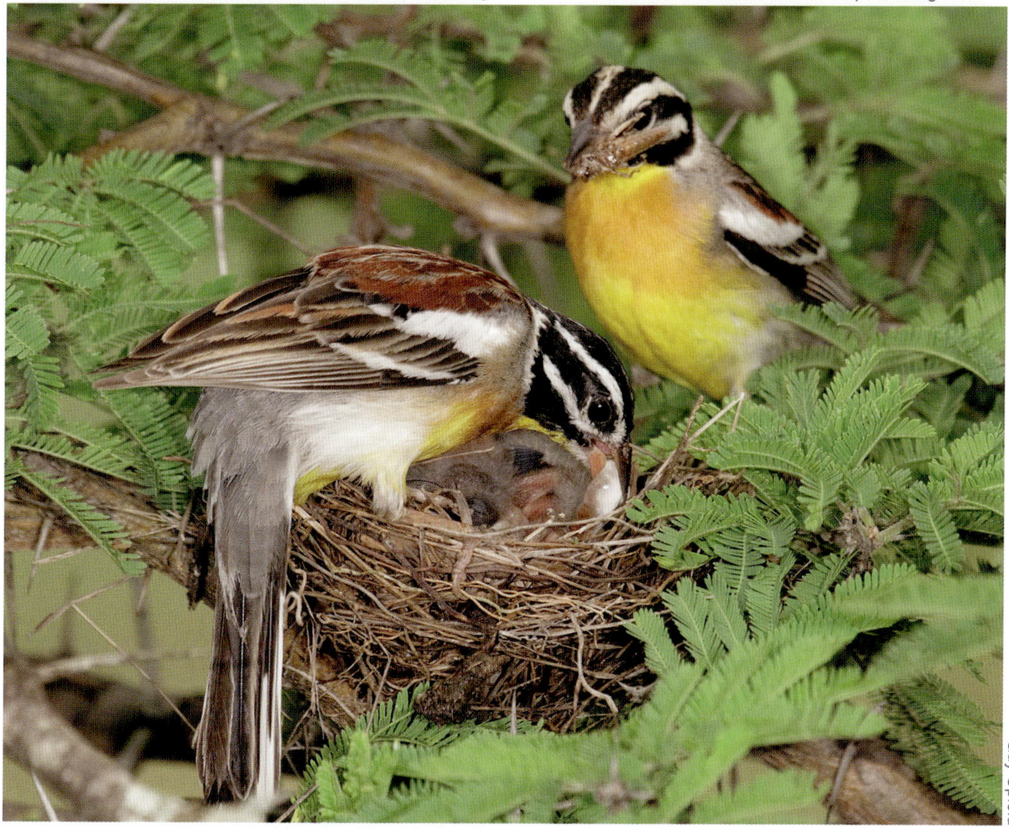

Guy Upfold

usokhandamidwombalabala Golden-breasted Bunting ♀ and ♂

GLOSSARY

A reminder that 'q.v.' (*quod vide*) means 'which see', a reference to another entry in this Glossary.

adjective: a part of speech (q.v.) which describes a noun (q.v.).

binomial: the two-part scientific names which consists of the genus name and the specific epithet.

biosystematics: the study of biological systems and the principles of classification and nomenclature.

cognates: nouns (q.v.) from different languages which have the same or similar shape and the same meanings, as a result of the languages all being derived from an earlier 'mother language'.

coinage: in this book, the creation of a new bird name out of existing elements in the language.

connotations: underlying meanings in certain nouns and names which suggest certain qualities: the word *owl* and names for owls in various languages may connote wisdom in some cultures, but connote witchcraft in others.

diachronic: of research or a particular study which takes into account data from an extended period of time. C.f. synchronic.

dialectal variations: the variations in word form (e.g. *ikhwinsi* vs *ikhwezi*) which result from different dialects, especially regional dialects.

dictionary trawling: going through dictionaries from cover to cover looking for specific data.

discontinuities: in a folk taxonomy, clusters (q.v) of birds are marked by what is similar to the birds in each cluster, and as well as what is different between the various clusters. In ethnobiology, these differences are known as 'discontinuities'.

epithet: in scientific nomenclature, the second half of a binomial (q.v.).

family: in scientific taxonomy, the level between Genus and Order. For example, all parrots, parakeets and lovebirds are in the Family *Psittacidae*.

folk generic: another way of saying 'cluster' in non-scientific categorization of birds.

folk taxonomy: the categorization and grouping of biological entities in a (usually preliterate) non-Western society.

Genus: in scientific taxonomy, the level between Family and Species. For example, the five species of nightjar found in KwaZulu-Natal all belong to the Genus *Caprimulgus*.

grammatical: describing the structure and rules of a language, ranging from word order in sentences to the different ways of forming verb tenses and singular and plural forms of nouns.

herald: as used in this book: to announce the onset of a season or time of year, or even time of day. For example "the cuckoo heralds the spring" = 'when the cuckoo calls, it is a sign that spring is here'.

hierarchy: a system of ordering things (animals, people, objects, etc) in graded levels, usually from the broadest category to the narrowest. In the scientific categorization of birds, they are ordered in a hierarchy with the broadest category Aves containing all birds, to the narrowest categories of species and sub-species.

honorific: as used in this book, a name that honors someone who has been important in the general field of ornithology: a collector of birds; a noted taxonomist; a noted author of ornithological works, etc. Honorifics may be found in scientific names (*Aquila wahlbergii*) or in vernacular names (Verreaux's Eagle Owl).

idiomatic: referring to the ways in which individual words are not taken at face value in a particular phrase (group of words) but together take on a new meaning. For example, in the phrase 'keep one's nose to the grindstone' neither an actual nose nor an actual grindstone are involved, and the idiomatic meaning of the phrase is 'work hard and continuously'.

lexical elements: The parts of words that appear separately in a dictionary. In the English compound noun *milkman*, the lexical elements are *milk* and *man*. In the Zulu compound noun *isigqobhamithi* (generic name for woodpeckers) the lexical elements are *gqobha* ('cut notches in') and *imithi* ('trees').

lexicographers: linguistic specialists who study the art of dictionary writing, or the writers of dictionaries.

GLOSSARY

linguist: as used in this book, an academic who studies various aspects of languages: their structure, their sounds, their history, their use in society and so on.

linguistics: the formal study of language.

literary devices: the different ways in which poets, dramatists and writers use and manipulate words for special effect.

metaphor: a literary device (q.v.) where one thing is compared to another but the comparative words are left out. Thus "Shaka is like a lion" is a simile (q.v.) because the words 'is like a' are present, but "Shaka, the Lion who caused ..." is a metaphor. The English vernacular bird name Trumpeter Hornbill is a metaphor.

morphological features: different aspects of the morphology (q.v.) of a word.

morphology (linguistics): the shape and structure of a word: its different parts and how these relate to each other.

morphology (biology): the shape and structure of a living entity (fish, plant, bird): its different parts and how these relate to one another.

nomenclature: a body of names, typically of a single group of entities (English-speaking peoples, plants, birds) and the system in which these names operate.

noun: the part of speech that typically functions as the subject and/or object of a verb in a sentence, as in "The *teacher* teaches the *children*".

omen: something that happens which in a traditional belief system indicates that something good, more usually evil, is to follow: "A broken mirror is an omen of seven years of bad luck".

onomastics: the academic discipline of studying names and naming systems.

onomatopoeia: a literary device (q.v.) where a word imitates a sound made in real life, as in *ka-boom!*, *zing!* and *plip-plop*.

pelagic: relating to deep sea (as opposed to 'coastal').

phonemes: units of speech sounds in a language.

phonological: referring to the phonology of a language: the speech sounds of a language and how they interact with one another.

portent: a synonym (q.v.) of omen. A sign that some good or bad (usually the latter) is about to happen. A Hamerkop landing on the roof of a hut is a portent of great evil to come.

praise poems: In southern Bantu societies, kings, chiefs, nobles and heroes are praised with oral poems which narrate their lineage, their great deeds, their chiefly attributes, and so on. The phenomenon is known as praise poetry; individual poems are praise poems.

prefixed: having a prefix (q.v.) added.

prefix: something added to the front of a word. In English *sub–*, *con–* and *mis–* in *submerge*, *contract* and *misunderstand* are all prefixes. Zulu verbs and nouns have a variety of possible prefixes and the name-forming prefixes *–no–*, *–so–*, *–sa–* and *–ma–* feature strongly in the formation of new bird names as listed in this book.

proverb: a popular saying or short sentence containing 'folk wisdom'. 'Make hay while the sun shines' is an English proverb which carries the 'deeper' wisdom of 'take your opportunities while you can'. Many Zulu proverbs feature birds, as does the English proverb 'One swallow does not make a summer'.

raptors: birds of prey

root forms: Most Zulu words have layers of prefixes and suffixes (q.v.). When these are removed, what is left is the root form.

salience: prominence or conspicuousness in the features of a bird. A bird may be visually salient (have striking or conspicuous features), be vocally salient (have a noticeable and individual call) or culturally salient: be a herald (q.v.) or carry omens and portents (q.v.).

semantic: relating to meaning. Semantics is the linguistic study of the meaning of words and layers and levels of meaning.

simile: a literary device (q.v.) where a comparison is made between one entity and another, and the comparison is explicit: 'Shaka is like a lion'. The Zulu prefix (q.v.) *–sa–* means 'just like', so bird names formed with this are using simile. See also *metaphor*.

GLOSSARY

suffix: an additional element placed after a word. Three common suffixes in English are *–less*, *–ful* and *–ness*, found in words like *helpful*, *helpless*, *rightful* and *rightfulness*. The most common suffix in Zulu bird names is *–ana*, indicating something small.

synonym (linguistics): any word in a group of two or more words which have the same meaning: *glad*, *happy*, *pleased*, *delighted* and *joyful* are all synonyms of one another.

synonym (biological): in scientific nomenclature (q.v.) earlier binomials (q.v.) which are no longer in use.

terminographic: the writing of terms in terminology development (q.v.).

terminology: (a) a body of words or terms in a specific language relating to a particular subject: 'the terminology of economics'; (b) the study of such specific terms.

terminology development: the creation of specialized terms in a specific language for specific subject areas. In this book (and in the Zulu Bird Name Project) 'terminology development' refers specifically to the creation of new Zulu names for birds.

transliteration: as used in this book, the term transliteration refers to the practice of adopting a word from another language into one's own. The Zulu language has a considerable number of adopted words from Afrikaans and English, such as *ibhulukwe* (Afrik. *broek* 'trousers') and *u-ayisikhilimu* (*ice-cream*). We have tried not to use this linguistic practice in the creation of new Zulu bird names.

vagrant: a bird which has strayed from its normal geographical location, often as a result of strong winds and storms.

vernacular: used specifically in this book to refer to any bird name which is not a scientific binomial (q.v.). All English and Zulu bird names in this book are regarded as vernacular names.

Distribution maps

A distribution map appears next to the species texts. The colour coding for these maps is provided in the colour key on the right. These maps are, in the main, a reflection of the data assembled by birders and 'citizen scientists' for the two Southern African Bird Atlas Projects (SABAP 1 and 2) and managed by the Animal Demography Unit (ADU), University of Cape Town. The maps should be interpreted only as guides to the general distribution and relative abundances of the species concerned. Out-of-range and vagrancy records are typically not reflected.

Colour key for distribution maps

REFERENCES

Alberts M. 2008. Translation-orientated terminography in the electronic age. Afrilex Conference Stellenbosch.

Bryant AT. 1905. *Zulu - English Dictionary*. Pinetown: Mariannhill Mission Press.

Chittenden H., G. Davies and I. Weiersbye. 2016. *Roberts Bird Guide*. Second Edition. Cape Town: John Voelcker Bird Book Fund.

Chittenden H., G. Davies and I. Weiersbye. 2018. *Roberts Voelgids. Tweede Uitgawe*. Cape Town: John Voelcker Bird Book Fund.

Doke CM. and BW Vilakazi. 1958. *Zulu-English Dictionary*. Johannesburg: Wits University Press.

Gill F, and M. Wright 2006. *Birds of the World: recommended English names*. Princeton, New Jersey and London: Princeton University Press.

Jewitt D., Goodman PS., Erasmus BFN., O'Connor TG., Witkowski ETF. 2015. Systematic land-cover change in KwaZulu-Natal, South Africa: implications for biodiversity. *South African Journal of Science* 111(9/10), 9 pgs.

Koopman A. 2018. Zulu bird names: A progression over the decades (I). *South African Journal of African Languages* 38(3): 261–267.

Koopman A. 2019 a. Zulu bird names: A progression over the decades (II). *South African Journal of African Languages* 39(1): 9–15.

Koopman A. 2019 b. *Zulu Bird Names and Bird Lore*. University of KwaZulu-Natal Press. Pietermaritzburg.

Koopman A., R. Porter, & N. Turner (2020). *Amagama Ezinyoni Zulu names of birds*. University of KwaZulu-Natal Press, Pietermaritzburg.

Marnewick MD., EF. Retief, NT. Theron, DR. Wright, TA. Anderson. 2015. *Important Bird and Biodiversity Areas of South Africa*. Birdlife South Africa, Johannesburg.

MacLean G. 1984. *Roberts' Birds of South Africa*. Fifth Edition. Cape Town: John Voelcker Bird Book Fund.

MacLean G. 1993. *Robert's Birds of South Africa*. Sixth Edition. Cape Town: John Voelcker Bird Book Fund.

Oliver J. 2003. *Ngezinyoni Zethu*. Translated from Oliver 1980 by D. Kumalo and Noleen Turner. Durban: WESSA and Birdlife South Africa.

Oliver J. 1980. *A Beginner's Guide to Our Birds*. Durban: Wildlife Society of Southern Africa.

Reedman R. 2016. *Lapwings, Loons and Lousy Jacks: The How and Why of Bird Names*. Exeter Pelagic Publishing.

Skerrett A., P. Matyot and G. Rocamora. 2003. Zwazo Sesel: The name of Seychelles Birds and their Meanings. Victoria, Seychelles: Island Conservation Society.

Taylor MR., F. Peacock & RM. Wanless (eds). 2015. *The Eskom Red Data Book of Birds of South Africa, Lesotho and Swaziland*. Birdlife South Africa, Johannesburg.

Tilman D., M. Clark, DR. Williams, K. Kimmel, S. Polasky & C. Packer. 2017. Future threats to biodiversity and pathways to their protection. *Nature* 546 pgs 73-81.

Turner N. & A. Koopman. 2018. Bird conservation in SA and terminology development. Presentation: 27th International Ornithological Congress. Vancouver, Canada.

NOTES

NOTES

NOTES

NOTES

INDEX A–C

A

abayeni **138**
Accipiter badius........................ 66
Accipiter melanoleucus 68
Accipiter minullus 66
Accipiter rufiventris 66
Accipiter tachiro....................... 66
Acridotheres tristis...................174
Acrocephalu baeticatus............162
Acrocephalus arundinaceus 160
Acrocephalus gracilorostris 160
Acrocephalus palustris 162
Acrocephalus schoenobaenus...160
Acrocephalus scirpaceus...........162
Actitis hypoleucus 92
Actophilornis africanus 90
Albatross 40
Alcedo semitorquata 124
Alopochen aegyptiaca...............34
Amadina erythrocephala..........200
amadojeyanabomvu **204**
amadojeyanajwayelekile **204**
amadojeyanalunga **204**
Amandava subflava 204
Amaureornis flavirostris 80
Amblyospiza albifrons 192
Anaplectes rubriceps................196
Anas capensis...........................36
Anas erythrorhyncha................. 38
Anas hottenttota....................... 38
Anas platyrhynchos 36
Anas smithii............................. 38
Anas sparsa..............................36
Anas undulata 36
Anastomus lamelligerus 42
Andropadus importunes...........152
Anhinga rufa 54
Anomalospiza imberbis 206
Anthoscopus caroli...................148
Anthreptes reichenowi 190
Anthus brachyurus...................210
Anthus caffer 210
Anthus chloris..........................212
Anthus cinnamomeus...............208
Anthus crenatus 212
Anthus hoeschi........................210
Anthus leucophrys...................212
Anthus lineiventris 210
Anthus nicholsoni 210
Anthus vaalensis 210
Apalis flavida 170
Apalis ruddi 170
Apalis thoracica 170
Apalis, Bar-throated 170
Apalis, Rudd's..........................170
Apalis, Yellow-breasted 170
Apaloderma narina 120
Apus affinis 118
Apus apus 118

Apus barbatus 118
Apus caffer..................................118
Apus horus 118
Apus melba 118
Aquila nipalensis..........................70
Aquila rapax................................ 68
Aquila spilogaster 70
Aquila verreauxii 70
Ardea alba 50
Ardea cinerea..............................52
Ardea goliath...............................52
Ardea melanocephala 52
Ardea pupurea 52
Ardeola ralloides 50
Arenaria interpes......................... 94
Asio capensis 112
Aviceda cuculoides 56
Avocet, Pied 84

B

Babbler, Arrow-marked............. 172
Balearica reguloram.................... 82
Barbet, Acacia Pied 132
Barbet, Black-collared 130
Barbet, Crested 130
Barbet, Green............................130
Barbet, White-eared 130
Bateleur..................................... 62
Batis capensis............................136
Batis fratrum.............................138
Batis molitor..............................138
Batis, Cape 136
Batis, Chinspot..........................138
Batis, Woodwards' 138
Bee-eater, European 126
Bee-eater, Blue-cheeked............126
Bee-eater, Little, ... 124
Bee-eater, Southern Carmine 126
Bee-eater, White-fronted............124
Bishop, Southern Red................198
Bishop, Yellow...........................198
Bishop, Yellow-crowned.............198
Bittern, Dwarf............................ 48
Bittern, Eurasian 48
Bittern, Little 48
Blackcap, Bush..........................172
Bokmakierie..............................140
Bostrychia hagedash................... 46
Botaurus stellaris 48
Boubou, Southern......................140
Bradypterus baboecala...............162
Bradypterus barratti 162
Broadbill, African 136
Brownbul, Terrestrial 154
Brubru......................................140
Bubo africanus 114
Bubo capensis 114
Bubo lacteus 114
Bubulcus ibis.............................. 50
Bucorvus leadbeateri 128

Bulbul, Dark-capped....................152
Bunting, Cape 216
Bunting, Cinnamon-breasted......216
Bunting, Golden-breasted 216
Buphagus erythrorhynchus..........176
Burhinus capensis 84
Burhinus vermiculatus 84
Bushshrike, Gorgeous..................140
Bushshrike, Grey-headed 138
Bushshrike, Olive........................138
Bushshrike, Orange-breasted......140
Bustard, Black-bellied 76
Bustard, Denham's 76
Bustard, White-bellied 76
Buteo buteo 68
Buteo rufofuscus.......................... 68
Buteo trizonatus 68
Butorides rufiventris....................52
Butorides striata 52
Buttonquail, Black-rumped..........32
Buttonquail, Common.................32
Buzzard, Common 68
Buzzard, European honey 56
Buzzard, Forest........................... 68
Buzzard, Jackal........................... 68
Buzzard, Lizard........................... 66
Bycanistes bucinator...................128

C

Calamonastes stierlingi 170
Calandrella cinerea....................150
Calendulauda sabota 150
Calidris alba 94
Calidris canutus 94
Calidris ferruginea....................... 94
Calidris melanotos 92
Calidris minuta 94
Camaroptera brachyuran 170
Camaroptera, Green-backed 170
Campephaga flava 142
Campethera abingoni 134
Campethera notata....................134
Campicoloides bifasciatus 182
Canary, Black-throated................212
Canary, Brimstone......................214
Canary, Cape.............................214
Canary, Forest............................212
Canary, Lemon-breasted 214
Canary, Yellow-fronted................214
Caprimulgus europaeus 116
Caprimulgus fossii 116
Caprimulgus natalensis 116
Caprimulgus pectoralis 116
Caprimulgus tristigma 116
Cecropsis abyssinica...................158
Cecropsis cucullata 158
Cecropsis semirufa.....................158
Centropus burchellii...................106
Centropus grillii.........................106
Cercotrichas leucophrys 182

226

INDEX C–D

Cercotrichas quadrivirgata........180
Cercotrichas signata...................180
Certhilauda semitorquata 150
Ceryle rudis 124
Ceuthmochares australis 108
Chaetops aurantius 144
Chalcomitra amethystina 188
Chalcomitra senegalensis 188
Charadrius hiaticula 88
Charadrius leschenaultia............. 88
Charadrius marginatus 88
Charadrius mongolus 88
Charadrius pecuarius 88
Charadrius tricollaris 88
Chat, Ant-eating...........................182
Chat, Buff-streaked......................182
Chat, Familiar..............................182
Chersomanes albofasciata...........150
Chlidonias hybrida.....................100
Chlidonias leucopterus................100
Chlorocichla flaviventris..............152
Chlorophoneus olivaceus 138
Chlorophoneus sulfureopectus ...140
Chroicocephalus cirrocephalus ... 96
Chrysococcyx caprius..................108
Chrysococcyx cupreus................. 110
Chrysococcyx klaas......................108
Ciconia abdimii 44
Ciconia ciconia 44
Ciconia episcopus 44
Ciconia nigra................................. 44
Cinnyricinclus leucogaster 174
Cinnyris afra 188
Cinnyris bifasciata 190
Cinnyris chalybeus 188
Cinnyris mariquensis..................190
Cinnyris neergaardi....................190
Cinnyris talatala 190
Circaetus cinereus 62
Circaetus fasciolatus 62
Circaetus pectoralis 62
Circus aeruginosus....................... 64
Circus macrourus......................... 64
Circus maurus 64
Circus pyargus 64
Circus ranivorus 64
Cisticola aberrans.......................164
Cisticola aridulus 168
Cisticola ayresii..........................168
Cisticola chiniana.......................166
Cisticola cinnamomeus...............168
Cisticola erythrops 164
Cisticola fulvicapilla...................164
Cisticola galactotes......................166
Cisticola juncidis 166
Cisticola lais 166
Cisticola natalensis 166
Cisticola textrix 168
Cisticola tinniens 166
Cisticola, Cloud...........................168

Cisticola, Croaking......................166
Cisticola, Desert..........................168
Cisticola, Lazy 164
Cisticola, Levaillant's..................166
Cisticola, Pale-crowned 168
Cisticola, Rattling 166
Cisticola, Red-faced....................164
Cisticola, Rufous-winged............166
Cisticola, Wailing 166
Cisticola, Wing-snapping 168
Cisticola, Zitting..........................166
Clamator glandarius 108
Clamator jacobinus 108
Clamator levaillanti....................108
Clanga pomarina 68
Cliff Chat, Mocking.....................184
Coccopygia melanotis 202
Colius striatus.............................120
Columba arquatrix......................102
Columba delegorguei 102
Columba guinea 102
Columba larvata 102
Columba livia..............................102
Coot, Red-knobbed......................82
Coracias caudata 120
Coracias garrulus 120
Coracias naevius 120
Coracina caesia 142
Cormorant, Cape 54
Cormorant, Reed..........................54
Cormorant, White-breasted 54
Corvus albicollis..........................148
Corvus albus 148
Corvus capensis..........................148
Corvus splendens 148
Corythaixoides concolor 106
Corythornis cristata....................124
Cossypha caffra...........................180
Cossypha dichroa........................180
Cossypha heuglini 178
Cossypha humeralis 180
Cossypha natalensis....................180
Coturnix coturnix 32
Coturnix delegorguei 32
Coucal, Black..............................106
Coucal, Burchell's 106
Courser, Bronze-winged............. 96
Courser, Temminck's................... 96
Crake, African 78
Crake, Baillon's............................. 80
Crake, Black 80
Crake, Corn...................................78
Crane, Blue 82
Crane, Grey Crowned.................. 82
Crane, Wattled 82
Creatophora cinerea 174
Crex crex......................................78
Crex egregia 78
Crithagra atrogularis 212
Crithagra citrinispectus..............214

Crithagra gularis 212
Crithagra mozambica.................214
Crithagra scotops........................212
Crithagra sulphurata..................214
Crithagra symonsi 214
Crombec, Long-billed..................160
Crow, Cape..................................148
Crow, House 148
Crow, Pied...................................148
Cuckoo Finch 206
Cuckoo, African Emerald 110
Cuckoo, African 110
Cuckoo, Black 110
Cuckoo, Common 110
Cuckoo, Diederik 108
Cuckoo, Great Spotted.................108
Cuckoo, Jacobin...........................108
Cuckoo, Klaas's............................108
Cuckoo, Levaillant's.....................108
Cuckoo, Red-chested 110
Cuckoo-Hawk, African 56
Cuckooshrike, Black 142
Cuckooshrike, Grey 142
Cuculus canorus 110
Cuculus clamosus........................ 110
Cuculus gularis........................... 110
Cuculus solitarius....................... 110
Cursorius temminckii................... 96
Cyanomitra olivacea...................188
Cyanomitra veroxii.....................188
Cypsiurus parvus 118

D

Dabchick 40
Darter, African..............................54
Delichon urbacum.......................158
Dendrocygna bicolour 34
Dendrocygna viduata 34
Dendroperdix sphaena................. 28
Dendropicos fuscescens 134
Dendropicos griseocephalus.......134
Dendropicos namaquus 134
Dicrurus adsimilis.......................146
Dicrurus ludwigii 146
Dove, Emerald-spotted................104
Dove, Laughing............................104
Dove, Lemon 102
Dove, Namaqua 104
Dove, Red-eyed............................104
Dove, Ring-necked.......................102
Dove, Rock...................................102
Dove, Tambourine 104
Dromas ardeola 84
Drongo, Fork-tailed......................146
Drongo, Southern Square-tailed.146
Dryoscopus cubla........................140
Duck, African Black.....................36
Duck, Fulvous Whistling 34
Duck, Knob-billed........................36
Duck, Maccoa...............................38

227

INDEX D–H

Duck, White-backed 34
Duck, White-faced Whistling 34
Duck, Yellow-billed 36

E

Eagle, African Fish 56
Eagle, Black-chested Snake 62
Eagle, Booted 72
Eagle, Brown Snake 62
Eagle, Crowned 70
Eagle, Lesser-spotted 68
Eagle, Long-crested 72
Eagle, Martial 70
Eagle, Southern Banded Snake 62
Eagle, Steppe 70
Eagle, Tawny 68
Eagle, Verreaux's 70
Eagle, Wahlberg's 72
Eagle-Owl, Verreaux's 114
Eagle-Owl, Cape 114
Eagle-Owl, Spotted 114
Egret, Great 50
Egret, Intermediate 50
Egret, Little 50
Egret, Western Cattle 50
Egretta ardesiaca *50*
Egretta garzetta *50*
Egretta intermedia *50*
Elanus caeruleus *58*
Emarginata familiaris *182*
Emberiza capensis *216*
Emberiza flaviventris *216*
Emberiza tahapisi *216*
Ephippiorhynchus senegalensis .. *44*
Eremomela icteropygialis *172*
Eremomela scotops *170*
Eremomela usticollis *172*
Eremomela, Burnt-necked 172
Eremomela, Green-capped 170
Eremomela, Yellow-bellied 172
Eremopterix leucotis *152*
Estrilda astrild *202*
Estrilda perreini *202*
Euplectes afer *198*
Euplectes albonotatus *200*
Euplectes ardens *198*
Euplectes axillaris *198*
Euplectes capensis *198*
Euplectes orix *198*
Euplectes progne *198*
Eupodotis caerulescens *76*
Eupodotis ruficrista *76*
Eupodotis senegalensis *76*
Eurystomus glaucurus *122*
Excalfactoria adansonii *32*

F

Falco amurensis *74*
Falco biarmicus *74*
Falco concolor *74*

Falco naumanni *72*
Falco peregrinus *74*
Falco rupicoloides *72*
Falco rupicolus *72*
Falco subbuteo *74*
Falco vespertinus *74*
Falcon, Amur 74
Falcon, Lanner 74
Falcon, Peregrine 74
Falcon, Red-footed 74
Falcon, Sooty 74
Finch, Red-headed 200
Finfoot, African 80
Firefinch, African 202
Firefinch, Jameson's 202
Firefinch, Red-billed 200
Fiscal, Southern 144
Flamingo, Greater 42
Flamingo, Lesser 42
Flufftail, Buff-spotted 78
Flufftail, Red-chested 78
Flufftail, Striped 78
Flufftail, White-winged 78
Flycatcher, African Dusky 186
Flycatcher, African Paradise 146
Flycatcher, Ashy 186
Flycatcher, Blue-mantled Crested .....
 146
Flycatcher, Fairy 146
Flycatcher, Fiscal 184
Flycatcher, Pale 184
Flycatcher, Southern Black 184
Flycatcher, Spotted 186
Francolin, Coqui 28
Francolin, Crested 28
Francolin, Grey-winged 30
Francolin, Red-winged 30
Francolin, Shelley's 30
Fulica cristata *82*

G

Gallinaga nigripennis *90*
Gallinula chloropus *80*
Gallinule, Allen's 82
Gannet, Cape 54
Geocolaptes olivaceus *134*
Geokichla gurneyi *176*
Geokichla guttata *178*
Geronticus calvus *46*
Glareola pratincola *96*
Glaucidium perlatum *114*
Go-away-bird, Grey 106
Godwit, Bar-tailed 90
Goose, African Pygmy 34
Goose, Egyptian 34
Goose, Spur-winged 34
Gorsachius leucotis *48*
Goshawk, African 66
Goshawk, Gabar 66
Grassbird, Cape 158

Grassbird, Fan-tailed 164
Grebe, Little 40
Greenbul, Sombre 152
Greenbul, Yellow-bellied 152
Greenbul, Yellow-streaked 154
Greenshank, Common 92
Ground Hornbill, Southern 128
Ground Thrush, Orange 176
Ground Thrush, Spotted 178
Grus carunculata *82*
Grus paradisea *82*
Guineafowl, Crested 28
Guineafowl, Helmeted 28
Gull, Grey-headed 96
Gull, Kelp 96
Guttera pucherani *28*
Gymnoris superciliaris *192*
Gypaetus barbatus *60*
Gypohierax angolensis *56*
Gyps africanus *60*
Gyps coprotheres *60*

H

Haematopus moquini *84*
Halcyon albiventris *122*
Halcyon chelicuti *122*
Halcyon senegalensis *122*
Halcyon senegaloides *122*
Haliaeetus vocifer *56*
Hamerkop 48
Harrier, African Marsh 64
Harrier, Black 64
Harrier, Montagu's 64
Harrier, Pallid 64
Harrier, Western Marsh 64
Harrier-Hawk, African 58
Hawk, Bat 56
Hawk-Eagle, African 70
Hawk-Eagle, Ayres's 70
Hedydipna collaris *190*
Helmetshrike, Retz's 138
Helmetshrike, White-crested 138
Heron, Black 50
Heron, Black-crowned Night 48
Heron, Black-headed 52
Heron, Goliath 52
Heron, Grey 52
Heron, Purple 52
Heron, Rufous-bellied 52
Heron, Squacco 50
Heron, Striated 52
Heron, White-backed Night 48
Hieraaetus ayresii *70*
Hieraaetus pennatus *72*
Hieraaetus wahlbergi *72*
Himantopus himantopus *84*
Hippolais icterina *164*
Hippolais olivetorum *164*
Hirundo albigularis *156*
Hirundo atrocaerulea *154*

INDEX H–I

Hirundo fuligula *156*
Hirundo rustica *156*
Hirundo smithii *158*
Hobby, Eurasian 74
Honeybird, Brown-backed 132
Honeyguide, Greater 132
Honeyguide, Lesser 132
Honeyguide, Scaly-throated 132
Hoopoe, African 126
Hoopoe, Green Wood 126
Hornbill, African Grey 128
Hornbill, Crowned 128
Hornbill, Southern Red-billed ... 128
Hornbill, Southern Yellow-billed 128
Hornbill, Trumpeter 128
Hydroprogne caspia *98*
Hypargos margaritatus *200*

I

ibhada 152
ibhaku lehlanze 206
ibhaku 198
ibhoboni 140
ibhoyi 170
Ibis, Glossy 46
Ibis, Hadada 46
Ibis, Sacred 46
Ibis, Southern Bald 46
ibomvana 198
idada laseYurobhu 36
idadelibomvu 36
idadelikhandamnyama 38
idadelimlomobomvu 38
idadelimlomophuzi 36
idadelimnyama 36
idadelincane 38
idlanga lentaba 60
idonsi 196
Iduna natalensis *162*
ifefelibomvu 122
ifefelihle 120
ifefeliluhlaza 120
ifefemidwa 120
ifubesi 114
igedezi 154
igelesha lehlathi 194
igelesha logu 194
igeleshelimqalonsundu 194
igwababa ledolobha 148
igwababa 148
igwababakazi 148
igwalagwala lehlanze 106
igwalagwala logu 106
igwalagwaleliluhlaza 106
igwedlamanzi 80
igwigwi 88
ihhahane 46
ihhoye 34
ihlabahlabane 124
ihlalankomo 176

ihlalanyathi 176
ihlathinyane 164
ihlekehle 172
ihlekehleke 172
ihlokohlokelikhulu 194
ihlokohloko lehlathi 196
ihlokohloko lomuzi 196
ihlokohlokwana 194
iholamvula 118
ihobhelimehlabomvu 104
ijankomo 118
ijikanyawo 58
ijiyankomelimlotha 118
ijuba ledolobha 102
ijubantondo 104
ijubelintamemhlophe 102
ikhonqwelo 142
ikhungula 42
ikhwelematsheni 184
ikhwelentabeni 182
ikhwezelimacwebi 174
Ikhwezi lasogwini 176
ikhwezi 176
ikhwikhwi 176
ikhwinsi laseYurobhu 174
ikhwinsi 176
iklebedwane 142
iklosi 148
ilanda 50
ilongwe 34
ilunga 144
imbove 136
imbuyelelo 122
imbuzane yomnqawe 172
imbuzane 170
imbuzaneluhlaza 170
imbuzanephuzi 172
imemela 128
impangele yehlathi 28
impangelejwayelekile 28
impevelimehlabomvu 138
impisiyolwandle 100
impofana 144
impofazana 174
impungayolwandle 40
imvukwane 40
imvuliyeza 118
imvumvuyane 66
incuncu 188
incuncwana 188
incwaba 138
incwinceluhlaza 188
incwincwemhlophe 190
incwincwemphungana 188
incwincwi yaseTembe 190
incwincwi yogu 190
incwincwi yomfomothi 190
indewula 96
indibilishi 160
Indicator indicator *132*

Indicator minor *132*
Indicator variegates *132*
Indigobird, Dusky 206
Indigobird, Village 204
indlangamandla 62
indlantuthwane 182
indlanyokemnyama 62
indlanyokempunga 62
indlanyokensundu 62
indlanyosi 126
indlazi 120
indudumela 86
indwe 82
ingcelekeshe 208
ingedana 132
ingede 132
ingeklencane 50
ingeklenkulu 50
ingongoni 140
ingophozi 104
ingqungqulu 62
ingududu 128
inguzambongolo 40
ingwangwa 176
inhlandlokazi 68
inhlangu 142
inhlavana 132
inhlavebizelayo 132
inhlekabafazi 126
inhlolamanzi 154
inhlolamfula 156
inhlolamvula yamadwala 156
inhlolamvula yasekhaya 158
inhlolamvulebhandensundu ....
.... 156
inhlolazulu 118
injubalukhalo 128
inkakulo 76
inkankane 46
inkankanelunga 46
inkankemidwa 108
inkanku 108
inkashana 200
inkonjane yamawa 156
inkonjane yaseYurobhu 156
inkonjanemnyama 154
inkonjanemqalomhlophe 156
inkonjanemqolomlotha 154
inkonjanencane 158
inkonjanenkulu 158
inkonjanesibhakabhaka 154
inkonjanesifubabomvu 158
inkonjanesileside 158
inkosiyezinkozi 70
inkotha 124
inkothana 124
inkothanyosi 126
inkothenkulu 126
inkovana 114
inkovu 140

229

INDEX I–L

inkukhumezane 80
inkukhuyamanzi 80
inkukhuyomhlanga 82
inkuletsheni 182
inkwali 30
inkwazana 56
inkwazi 56
inqe lehlanze 60
inqelendlovu 44
inqelincane 60
inqemvuma 56
inqomfebomvu 208
inqomfi 208
inqondanqonda 134
inqwathane 190
insansa 178
insansane 178
insewane 202
insingizi 128
insonsi 190
insukakude 100
insusha 188
inswempe 28
inswinswi 176
intaka 198
intakajolwane 152
intakansinsi 198
intakanyosi 198
intakemaphikamhlophe 200
intengu 146
intengwana 146
intingamafu 168
intinganqi 168
intingelilayo 166
intinginono 58
intiyanejwayelekile 202
intiyaneluhlaza 200
intonso 190
intshe 28
intungunono 58
intunjana 130
intuntwane 82
inxenge 204
inyoninyoka 54
inzwece 146
inzwinzwebomvu 34
inzwinzwi 34
inzwinzwinzwi 34
iphishamanzi 54
iphothwe 152
iqobo 166
iqola 144
iqotshana 168
isabhonsi 204
isadube 170
isagqukwe 102
isagundwane 184
isagwacesibhakabhaka 32
isagwacesibomvu 32
isagwacesijwayelekile 32

isakabuli 198
isandondondwane 130
isangulube 140
isankawu 38
isanqunzi 150
isanxa 68
isaqola 184
iseme 76
isicagogwane 160
isicelankobe 202
isichegu 182
isicibamanzi 54
isicibilili 196
isicivo 140
isicukujeje 148
isigqobhamithesiluhlaza ... 134
isigqobhamithi saseningizimu 134
isigqobhamithi 134
isigqobhamithintshebe 134
isigqobhamnenke 42
isigwe 198
isihlalamatsheni 184
isihuhwa 70
isikamanzi 40
isikhombazane sehlanze 104
isikhombazane sehlathi 104
isikhova sexhaphozi 112
isikhova sotshani 112
isikhovampondo 114
isikhovamponjwana 114
isikhovanhlanzi 114
isikhwayimba 142
isikhwehlesimqalabomvu ... 30
isikhwehlesimqhalomhlophe 30
isikhwehlesiqhova 28
isikhwelemaweni 184
isikhwenene 136
isikhwenenesikhandansundu 136
isikilothi 122
isinqonqotho 130
isiphikeleli 122
isiphungumangathi 72
isipopopo 130
isiqhanazana 80
isithandamanzi 44
isivuba 124
isiwelewele 84
isixula 124
isixulamasele 46
isizinzana 80
isizinzi 80
isomi 176
Ispidina picta 122
ithendelelibomvu 30
ithendelelimlotha 30
ithimbakazane 138
ititihoye 86

ititihoyenomqhele 86
itoyiya 208
ivevenyane 34
ivubelikhulu 42
ivubelincane 42
ivukuthu lehlathi 102
ivukuthu 102
ivuzigazi 202
iwabayi 148
iwamba 96
iwonde lasolwandle 54
iwondelimhlophe 54
Ixobrychus minutus 48
Ixobrychus sturmii 48

J

Jacana, African 90
Jacana, Lesser 90
Jynx ruficollis 132

K

Kaupifalco monogrammicus 66
Kestrel, Greater 72
Kestrel, Lesser 72
Kestrel, Rock 72
Kingfisher, African Pygmy ... 122
Kingfisher, Brown-hooded .. 122
Kingfisher, Giant 124
Kingfisher, Half-collared ... 124
Kingfisher, Malachite 124
Kingfisher, Mangrove 122
Kingfisher, Pied 124
Kingfisher, Striped 122
Kingfisher, Woodland 122
Kite, Black 58
Kite, Black-winged 58
Kite, Yellow-billed 58
Knot, Red 94
Korhaan, Blue 76
Korhaan, Red-crested 76

L

Lagonosticta rhodopareia .. 202
Lagonosticta rubricata 202
Lagonosticta senegala 200
Lamprotornis bicolour 176
Lamprotornis nitens 176
Laniarius ferrugineus 140
Lanius collaris 144
Lanius collurio 144
Lanius minor 144
Lapwing, African Wattled ... 86
Lapwing, Blacksmith 86
Lapwing, Black-winged 86
Lapwing, Crowned 86
Lapwing, Senegal 86
Lark, Eastern Long-billed .. 150
Lark, Flappert 150
Lark, Pink-billed 152
Lark, Red-capped 150

INDEX L–P

Lark, Rufous-naped 150
Lark, Sabota 150
Lark, Spike-heeled 150
Larus dominicanus 96
Leptoptilos crumenifer 44
Limosa lapponica 90
Lioptilus nigricapillus 172
Lissotis melanogaster 76
Lonchura cucullata 204
Lonchura fringilloides 204
Lonchura nigriceps 204
Longclaw, Cape 208
Longclaw, Rosy-throated 208
Longclaw, Yellow-throated 208
Lophaetus occipitalis 72
Lophoceros alboterminatus 128
Lophoceros nasutus 128
Lybius leucomelas 132
Lybius torquatus 130

M

Macheiramphus alcinu 56
Macronyx ameliae 208
Macronyx capensis 208
Macronyx croceus 208
Malaconotus blanchoti 138
Malkoha, Green 108
Mallard 36
Mandingoa nitidula 200
Mannikin, Bronze 204
Mannikin, Magpie 204
Mannikin, Red-backed 204
Martin , Banded 156
Martin, Brown-throated 154
Martin, Common House 158
Martin, Rock 156
Martin, Sand 156
Megaceryle maxima 124
Melaenornis pallidus 184
Melaenornis pammelaina 184
Melaenornis silens 184
Melaniparus niger 148
Merops apiaster 126
Merops bullockoides 126
Merops nubicoides 126
Merops persicus 126
Merops pusillus 124
Microcarbo africanus 54
Micronisus gabar 66
Microparra capensis 90
Milvus aegyptius 58
Milvus migrans 58
Mirafra Africana 150
Mirafra rufocinnamomea 150
Monticola explorator 184
Monticola rupestris 184
Moorhen, Common 80
Moorhen, Lesser 80
Morus capensis 54
Motacilla aguimp 208

Motacilla capensis 206
Motacilla clara 208
Motacilla flava 206
Mousebird, Red-faced 120
Mousebird, Speckled 120
Muscicapa adusta 186
Muscicapa caerulescens 186
Muscicapa striata 186
Mycteria ibis 42
Myioparus plumbeus 186
Myna, Common 174
Myrmecocichla formicivora 182
Myrmecocihla monticola 182

N

Necrosyrtes monachus 60
Nectarinia famosa 186
Neddicky 164
Neophron percnopterus 60
Neotis denhami 76
Netta erythrophthalma 38
Nettapus auritus 34
Nicator gularis 152
Nicator, Eastern 152
Nightjar, European 116
Nightjar, Fiery-necked 116
Nightjar, freckled 116
Nightjar, Square-tailed 116
Nightjar, Swamp 116
Nilaus afer 140
Notopholia corruscus 176
Numenius phaeopus 90
Numida meleagris 28
Nycticorax nycticorax 48

O

Oena capensis 104
ojojekhaya **206**
oklebeklebe **74**
Onychognathus morio 176
Onychoprion fuscatus 98
Openbill, African 42
Oriole, Black-headed 144
Oriole, Eurasian Golden 144
Oriolus larvatus 144
Oriolus oriolus 144
Ortygospiza atricollis 204
Osprey, Western 56
Ostrich, Common 28
Otus senegalensis 112
Owl, African Grass 112
Owl, African Scops 112
Owl, African Wood 112
Owl, Marsh 112
Owl, Pel's Fishing 114
Owl, Southern White-faced 114
Owl, Western Barn 112
Owlet, Pearl-spotted 114
Oxpecker, Red-billed 176
Oxyura maccaoa 38

Oystercatcher, African 84

P

Pandion haliaetus 56
Paragallinula angulata 80
Parakeet, Rose-ringed 136
Parrot, Brown-headed 136
Parrot, Cape 136
Passer diffuses 192
Passer domesticus 192
Passer melanurus 192
Pavo cristatus 28
Peafowl, Indian 28
Pelecanus onocrotalus 42
Pelecanus rufescens 42
Pelican, Great White 42
Pelican, Pink-backed 42
Peliperdix coqui 28
Penguin, African 40
Pernis apivorus 56
Petrel 40
Petronia, Yellow-throated 192
Phalacrocorax capensis 54
Phalacrocorax lucidus 54
Philomachus pugnax 96
Phoeniconaias minor 42
Phoenicopterus roseus 42
Phoeniculus cyanomelas 126
Phoeniculus purpureus 126
Phyllastrephus terrestris 154
Phyllastrepus flavostriatus 154
Phylloscopus ruficapilla 160
Phylloscopus trochilus 160
Pigeon, African Green 104
Pigeon, African Olive 102
Pigeon, Eastern Bronze-naped ... 102
Pigeon, Speckled 102
Pipit, African Rock 212
Pipit, African 208
Pipit, Buffy 210
Pipit, Bushveld 210
Pipit, Mountain 210
Pipit, Nicholson's 210
Pipit, Plain-backed 212
Pipit, Short-tailed 210
Pipit, Striped 210
Pipit, Yellow-breasted 212
Platalea alba 46
Platysteira peltata 136
Plectropterus gambensis 34
Plegadis falcinellus 46
Ploceus bicolour 196
Ploceus capensis 194
Ploceus cucullatus 196
Ploceus intermedius 194
Ploceus ocularis 192
Ploceus subaureus 194
Ploceus velatus 194
Ploceus xanthops 194
Ploceus xanthopterus 194

INDEX P–S

Plover, Common Ringed 88
Plover, Crab 84
Plover, Greater Sand 88
Plover, Grey 86
Plover, Kittlitz's 88
Plover, Lesser Sand 88
Plover, Three-banded 88
Plover, White-fronted 88
Pluvialis squatarola 86
Pochard, Southern 38
Podica senegalensis 80
Pogoniulus bilineatus 130
Pogoniulus pusillus 130
Pogonocichla stellata 178
Poicephalus cryptoxanthus 136
Poicephalus robustus 136
Polemaetus bellicosus 70
Polyboroides typus 58
Porphyria alleni 82
Porphyrio madagascariensis 82
Porzana pusilla 80
Pratincole, Collared 96
Prinia hypoxantha 168
Prinia subflava 168
Prinia, Drakensberg 168
Prinia, Tawny-flanked 168
Prion 40
Prionops plumatus 138
Prionops retzii 138
Prodotiscus regulus 132
Promerops gurneyi 186
Psalidoprocne holomelas 154
Pseudhirundo griseopyga 154
Psittacula krameri 136
Pternistis afer 30
Pternistis natalensis 30
Pternistis swainsonii 30
Pterochelidon spilodera 156
Ptilopsis granti 114
Puffback, Black-backed 140
Pycnonotus tricolour 152
Pytilia melba 200
Pytilia, Green-winged 200

Q

Quail, Blue 32
Quail, Common 32
Quail, Harlequin 32
Qualifinch 204
Quelea erythrops 196
Quelea quelea 196
Quelea, Red-billed 196
Quelea, Red-headed 196

R

Rail, African 80
Rallus caerulescens 80
Raven, White-necked 148
Recurvirostra avosetta 84
Redshank, Common 92

Reeve 96
Rhinoptilus chalcopterus 96
Riparia cincta 156
Riparia paludicola 154
Riparia riparia 156
Robin, White-starred 178
Robin-Chat, Cape 180
Robin-Chat, Chorister 180
Robin-Chat, Red-capped 180
Robin-Chat, White-browed 178
Robin-Chat, White-throated 180
Rock Thrush, Cape 184
Rock Thrush, Sentinel 184
Rockjumper, Drakensberg 144
Roller, Broad-billed 122
Roller, European 120
Roller, Lilac-breasted 120
Roller, Purple 120
Rostratula benghalensis 90
Ruff 96

S

Sagittarius serpentarius 58
Sanderling 94
Sandpiper, Common 92
Sandpiper, Curlew 94
Sandpiper, Marsh 92
Sandpiper, Pectoral 92
Sandpiper, Terek 94
Sandpiper, Wood 92
Sarkidiornis melanotos 36
Sarothrura affinis 78
Sarothrura ayresii 78
Sarothrura elegans 78
Sarothrura rufa 78
Saw-wing, Black 154
Saxicola torquatus 182
Schoenicola brevirostris 164
Scimitarbill, Common 126
Scleroptila afra 30
Scleroptila levaillantii 30
Scleroptila shelleyi 30
Scopus umbretta 48
Scotopelia peli 114
Scrub Robin, Bearded 180
Scrub Robin, Brown 180
Scrub Robin, White-browed 182
Secretarybird 58
Seedeater, Streaky-headed 212
Serinus canicollis 214
Shearwater 40
Shelduck, South African 36
Shikra 66
Shoveler, Cape 38
Shrike, Lesser Grey 144
Shrike, Magpie 142
Shrike, Red-backed 144
Siskin, Drakensberg 214
Skua, Brown 100
Smithornis capensis 136

Snipe, African 90
Snipe, Greater Painted 90
Sparrow, Cape 192
Sparrow, House 192
Sparrow, Southern Grey-headed . 192
Sparrowhawk, Black 68
Sparrowhawk, Little 66
Sparrowhawk, Rufous-breasted ... 66
Sparrow-Lark, Chestnut-backed . 152
Spheniscus demersus 40
Sphenoeacus afer 158
Spizocorys conirostris 152
Spoonbill, African 46
Spurfowl, Natal 30
Spurfowl, Red-necked 30
Spurfowl, Swainson's 30
Stactolaema leucotis 130
Stactolaema olivacea 130
Starling, Black-bellied 176
Starling, Cape 176
Starling, Common 174
Starling, Pied 176
Starling, Red-winged 176
Starling, Violet-backed 174
Starling, Wattled 174
Stenostira scita 146
Stephanoaetus coronatus 70
Stercorarius antarcticus 100
Sterna hirundo 100
Sterna paradisaea 100
Sternula albifrons 98
Stilt, Black-winged 84
Stint, Little 94
Stonechat, African 182
Stork, Abdim's 44
Stork, Black 44
Stork, Saddle-billed 44
Stork, White 44
Stork, Woolly-necked 44
Stork, Yellow-billed 42
Stork. Marabou 44
Storm Petrel 40
Streptopelia capicola 102
Streptopelia semitorquata 104
Streptopelia senegalensis 104
Strix woodfordii 112
Struthio camelus 28
Sturnus vulgaris 174
Sugarbird, Gurney's 186
Sunbird Amethyst 188
Sunbird, Collared 190
Sunbird, Greater Double-collared.....
....188
Sunbird, Grey 188
Sunbird, Malachite 186
Sunbird, Marico 190
Sunbird, Neergard's 190
Sunbird, Olive 188
Sunbird, Plain-backed 190
Sunbird, Purple-banded 190

INDEX S–U

Sunbird, Scarlet-chested 188
Sunbird, Southern Double-collared.. 188
Sunbird, White-bellied 190
Swallow, Barn 156
Swallow, Blue 154
Swallow, Greater Striped 158
Swallow, Grey-rumped 154
Swallow, Lesser Striped 158
Swallow, Red-breasted 158
Swallow, South African Cliff 156
Swallow, White-throated 156
Swallow, Wire-tailed 158
Swamphen, African 82
Swift, African Black 118
Swift, African Palm 118
Swift, Alpine 118
Swift, Common 118
Swift, Horus 118
Swift, Little 118
Swift, White-rumped 118
Sylvia borin *172*
Sylvia subcaerulea *172*
Sylvietta rufescens *160*

T

Tachibaptus ruficollis.................... *40*
Tadorna cana *36*
Tauraco corythaix..........................*106*
Tauraco livingstonii.......................*106*
Tauraco porphyreolophus...........*106*
Tchagra australis...........................*142*
Tchagra senegalus *142*
Tchagra tchagra *142*
Tchagra, Black-crowned 142
Tchagra, Brown-crowned 142
Tchagra, Southern 142
Teal, Cape 36
Teal, Hottentot................................ 38
Teal, Red-billed............................... 38
Telephorus viridis *140*
Telephorus zeylonus.....................*140*
Terathopius ecaudatus 62
Tern, Arctic 100
Tern, Caspian 98
Tern, Common 100
Tern, Lesser-crested 98
Tern, Little 98
Tern, Sandwich 98
Tern, Sooty 98
Tern, Swift 98
Tern, Whiskered 100
Tern, White-winged 100
Terpsiphone viridis *146*
Thalasseus bengalensis.................. 98
Thalasseus bergii 98
Thalasseus sandvicensis 98
Thalassornis leuconotus *34*
*Thamnolaea cinnamomeiventris*184
Thick-knee, Spotted...................... 84

Thick-knee, Water 84
Threskiornis aethiopicus............... *46*
Thrush, Groundscraper 178
Thrush, Kurrichane 178
Thrush, Olive 178
Tinkerbird, Red-fronted 130
Tinkerbird, Yedllow-rumped.......130
Tit, Grey Penduline......................148
Tit, Southern Black 148
Tit-Flycatcher, Grey......................186
Tockus erythrorhynchus *128*
Tockus leucomelas.......................*128*
Torgos tracheiotos *62*
Trachyphonus vaillantii *130*
Treron calvus *104*
Trigonoceps occipitalis *62*
Tringa glareola *92*
Tringa nebularia *92*
Tringa stagnatilis............................ *92*
Tringa totanus *92*
Trochocerus cyanomelas *146*
Trogon, Narina 120
Turaco, Knysna 106
Turaco, Livingstone's...................106
Turaco, Purple-crested 106
Turdoides jardineii........................*172*
Turdus libonyana *178*
Turdus litsitsirupa *178*
Turdus olivaceus *178*
Turnix nanus *32*
Turnix sylvaticus............................ *32*
Turnstone, Ruddy.......................... 94
Turtur chalcospilos *104*
Turtur tympanistra *104*
Twinspot, Green 200
Twinspot, Pink-throated 200
Tyto alba.. *112*
Tyto capensis................................. *112*

U

u(lu)ve 146
ubantwanyana 110
ubhaklakliyo 98
ubhamukwe 82
ubhavuzile wehlathi 78
ubhavuzilobomvana 78
ubhavuzilomhlophe 78
ubhavuzilomidwayidwa 78
ubucubu 202
ubusukuswane 202
ucijomhlophe 92
udemezane 58
udokotela 136
uduku 84
ufukwe 106
ufukwomnyama 106
ufumba 76
ugaga 180
ugaganomidwa 182
ugaganonsundu 180

ugaganontshebe 180
ugazini 200
uheshanobomvu 66
uheshanomnyama 68
uheshanyana 66
uheshomidwayidwa 66
uheshomlotha 66
uhlakayiya 146
uhlazazana 186
uhlekwane 206
ujamelamafuku 162
ujamelumhlanga wasemzansi 162
ujamelumhlanga wasenyakatho 162
ujamelumhlanga waseYurobhu 162
ujamelumhlangomkhulu.... 160
ujamelumhlangomncane ... 160
ujamelumhlangophuzi 162
ujenga wokhahlamba 168
ujenga 168
ujojo 198
ujolwane wehlanze 192
ujolwane wekhaya 192
ujolwanokhandaphunga 192
ukhandabomvu 196
ukhandaklebhu 196
ukhandelimhlophe 62
ukholwasomkhulu 42
ukholwasomncane 42
ukhonzane 104
ukhozilwentshebe 60
ukhozimuhlwa 70
ukhoziqholwane 70
ukhozolumabala 68
ukhozolumabalabala 70
ukhozolumadladla 72
ukhozolumidwayidwa 70
ukhozolumnyama 70
ukhozolunsundu 68
ukhozolusisila 72
uklebe lwehlathi 68
uklebemawa 74
uklebompunga 74
uklebonyawobomvu 74
uklebosankonjane 74
uklebosikhweshekweshe 74
ukliyo 98
umabhashinhlayela 140
umabhashinhlayelohlaza ... 138
umabhengwane 112
umabhocashile 48
umabhumfashane 48
umabilwane 170
umachibini 100
umacibudaka 46
umacuthobomvu 48
umacutholuhlaza 52
umacuthomhlophe 50

233

INDEX U

umacuthomncane 48	umbhalanonsundu 212	undlunkulu 192
umacuthomnyama 52	umbhekle 180	undodosibona 110
umadevaphuzi 86	umbhukwane 76	ungcelekeshephuzi 212
umadletshana 112	umbicini 174	ungolwane 32
umadolabomvu 44	umcwiewicwi 108	ungoqo 32
umagevuzomkhulu 88	umdokwe 166	ungozwana 94
umagevuzomncane 88	umehlwanoluhlaza 174	ungqangendlela 150
umagumejana 200	umehlwanophuzi 174	unhloyile waseYurobhu 58
umahambehlala 86	umfelokazomlomobomvu .. 204	unhloyile 58
umahlombamhlophe 44	umfelokazomlomomhlope . 206	unkombose 104
umahlwithilulwane 56	umgugwane 106	unkovuka 130
umakhalelilanga 118	umhlohlongwane 116	unobathekeli 112
umakhwaneni 90	umjekejeke waseAfrika 78	unobulongwana 96
umakhwaphamnyama 86	umjekejeke wasenyakatho ... 78	unobulongwonsundu 96
umakhwifikwifi 92	umjenenengu 120	unochibi 100
umananda 214	umjombo 120	unochweba 96
umalusinkomo 152	umkhololwane 128	unocu 136
umalwelwe 116	umkholompunga 128	unodaka 90
umambathilanga 198	umkholwane 128	unofosholo 38
umambathingubo 126	umkholwanomlotha 128	unogilomnyama 212
umamhlangeni wasentshon-alanga 64	umklewu 106	unogilonkomnyama 50
umamhlangenomnyama 64	umlindankomo 50	unogolantethe 44
umamhlangenomphunga 64	umloyi 112	unogozwana 94
umamhlangenonsundu 64	ummbesi 184	unogqabakazi 96
umananda 180	umncgelu wasematsheni ... 212	unogqabokhwifi 92
umandlankala 84	umncishu 206	unogxumetsheni 144
umandubulu 114	umngcelohlanze 210	unohemu 82
umangube 198	umngcelomidwa 210	unohhalulwandle 40
umankole 170	umngceloze 212	unohlohlweni 64
umankolophuzi 170	umngcelu wegquma 210	unokhifi 90
umantantolwandle 40	umngcelu wehlungu 210	unokhoboyi 52
umantuluza 186	umngcelu wentaba 210	unokhukhuza 110
umanyatheludaka 50	umngcelwana 210	unokilonkojwayelekile 52
umanyovini 56	umngqithi 76	unokilonkolikhandamnyama 52
umaphendulamatshe 94	umngquphane 142	unolunga 144
umaphithizela 94	umnqangqandolo 134	unolwandle 98
umashiyabomvu 136	umnqube wogu 138	unomanduli 160
umashwili 152	umnqumo 172	unomemeza 30
umashwilomidwa 154	umntoli 150	unomhlangomncane 82
umasikulufu 136	umqalaphuzi 160	unomlenzobomvu 92
umatatazela 88	umqonqotho 142	unomlomophuzi 42
umathandaluziba 90	umqoqongo 144	unomnqumo 164
umathantatha 90	umqwayini 66	unompempe 92
umathebethebana wamadwala. 72	umtshivovo 120	unomtsheketshe 132
umathebethebanomkhulu 72	umunswi wehlathi 178	unomtshingo 178
umathebethebanomncane ... 72	umunswili 178	unomunga 132
umathithibala 144	umvemventaba 208	unondindwa 206
umatsheni 116	umvemvolunga 208	unondwayizomkhulu 90
umavelashone 162	umvemvophuzi 206	unondwayizomncane 90
umawewe 116	umxwagele 46	unonengekhanda 108
umazabelweni 58	umzolozolo 126	unongilobomvu 132
umazalashiye 108	umzwelele 112	unongoyana 130
umbangaqhwa 84	umzwilili 214	unongozolwane 122
umbexe 182	umzwingili 194	unongqwashi 152
umbhalane wehlathi 212	unanaza 152	unonkalankala 122
umbhalane wokhahlamba . 214	uncede wehlane 168	unonklilwane 98
umbhalanomadevu 214	uncede 164	unonkliyo 98
umbhalanomkhulu 214	uncedobomvu 164	unonkliyomnyama 98
	uncedomnyama 166	unonklwe 204
	uncedoselesele 166	

INDEX U–Z

unonkositini 180
unonqane 118
unontenteza 100
unontshiloza 158
unonzwili 150
unopheshwana 94
unosigqokomnyama 172
unosikhutha 36
unosimila 36
unosiqalaba 186
unosongo 88
unosungulo 126
unothwayiza 92
unothwayizana 94
unotoyobomvu 208
unovilane 164
unovimba 94
unowanga 44
unozalashiyomabala 108
unozalizingwenya 52
unozila 84
unozulane 40
unqothi 150
untilontilo 150
unukani 126
unununde 90
uphalane 60
uphezukomkhono 110
upigogo 28
Upupa Africana 126
uqaqashe 150
uqhelu 166
uqholompunga 146
uqholwane 146
Uraeginthus angolensis 202
Urocolius indicus 120
Urolestes melanoleucus 142
usacingo 140
usagwebe 134
usakhukhuza 110
usamdokwe 102
usantiyane 200
ushowe 142
usiba 48
usibagwebe 130
usibó 144
usifubabomvu 188
usikhothambuzane 186
usikhothamlotha 186
usikhothaphela 164
usipheshula 84
usipoki 138
usiqophokezi 192
usiqhovana 130
usokhandamidwa wamatshe ....
.... 216
usokhandamidwombalabala ....
.... 216
usokhandamidwonsundu ... 216
usomheshe 66
usomthende 56
usonambuzane 186
usonkanyezi 178
usothathizwe 174
usotshanini 164
usozi 66
uswenka 124
uthekwane 48
uwili 152
uzangozolo 124
uzavolo 116
uzazu 128
uzibukwana 192

V

Vanellus armatus 86
Vanellus coronatus 86
Vanellus lugubris 86
Vanellus melanopterus 86
Vanellus senegallus 86
Vidua chalybeate 204
Vidua funerea 206
Vidua macroura 206
Vidua paradisaea 206
Vulture, Bearded 60
Vulture, Cape 60
Vulture, Egyptian 60
Vulture, Hooded 60
Vulture, Lappet-faced 62
Vulture, Palm-nut 56
Vulture, White-backed 60
Vulture, White-headed 62

W

Wagtail, African Pied 208
Wagtail, Cape 206
Wagtail, Mountain 208
Wagtail, Western Yellow 206
Warbler, African Reed 162
Warbler, African Yellow 162
Warbler, Barratt's 162
Warbler, Chestnut-vented .. 172
Warbler, Eurasian Reed 162
Warbler, Garden 172
Warbler, Great Reed 160
Warbler, Icterine 164
Warbler, Lesser Swamp 160
Warbler, Little Rush 162
Warbler, Marsh 162
Warbler, Olive-tree 164
Warbler, Sedge 160
Warbler, Willow 160
Warbler, Yellow-throated Woodland
.... 160
Wattle-eye, Black-throated .. 136
Waxbill, Blue 202
Waxbill, Common 202
Waxbill, Grey 202
Waxbill, Orange-breasted .. 204
Waxbill, Swee 202
Weaver, Cape 194
Weaver, Dark-backed 196
Weaver, Eastern Golden 194
Weaver, Holub's Golden 194
Weaver, Lesser Masked 194
Weaver, Red-headed 196
Weaver, Southern Brown-throated ...
.... 194
Weaver, Southern Masked ... 194
Weaver, Spectacled 192
Weaver, Thick-billed 192
Weaver, Village 196
Wheatear, Mountain 182
Whimbrel 90
White-eye, Cape 174
White-eye, Southern Yellow ... 174
Whydah, Long-tailed Paradise ... 206
Whydah, Pin-tailed 206
Widowbird, Fan-tailed 198
Widowbird, Long-tailed 198
Widowbird, Red-collared .. 198
Widowbird, White-winged .. 200
Woodpecker, Bearded 134
Woodpecker, Cardinal 134
Woodpecker, Golden-tailed .. 134
Woodpecker, Ground 134
Woodpecker, Knysna 134
Woodpecker, Olive 134
Wren-Warbler, Stierling's .. 170
Wryneck, Red-throated 132

X

Xenus cinereus 94

Z

Zosterops anderssoni 174
Zosterops virens 174

QUICK INDEX

Apalis 170	Finch 200	Moorhen 80	Shelduck................36
Avocet 84	Finfoot.................. 80	Mousebird............120	Shikra 66
Babbler................. 172	Firefinch............... 200	Myna 174	Shoveler 38
Barbet.................. 130	Fiscal 144	Neddicky 164	Shrike 142
Bateleur................. 62	Flamingo 42	Nicator 152	Siskin................... 214
Batis 136	Flufftail...................78	Nightjar 116	Skua 100
Bee-eater 124	Flycatcher 184	Openbill 42	Snake Eagle........... 62
Bishop 198	Flycatcher, African	Oriole 144	Snipe 90
Bittern 48	Paradise 146	Osprey 56	Sparrow............... 192
Blackcap 172	Francolin............... 28	Ostrich 28	Sparrowhawk......... 66
Bokmakierie...........140	Gallinule............... 82	Owl 112	Sparrow-Lark........ 152
Boubou 140	Gannet...................54	Owlet 114	Spoonbill 46
Broadbill 136	Go-away-bird 106	Oxpecker 176	Spurfowl 30
Brownbul 154	Godwit 90	Oystercatcher......... 84	Starling................ 174
Brubru..................140	Goose 34	Parakeet 136	Stilt....................... 84
Bulbul.................. 152	Goshawk............... 66	Parrot 136	Stint...................... 94
Bunting 216	Grassbird 158,164	Peafowl................. 28	Stonechat 182
Bushshrike 138	Grebe.................... 40	Pelican...................42	Stork..................... 44
Bustard...................76	Greenbul 152	Penguin 40	Storm Petrel........... 40
Buttonquail 32	Greenshank 92	Petrel 40	Sugarbird 186
Buzzard 68	Guineafowl 28	Petronia 192	Sunbird 186
Buzzard,Lizard........ 66	Gull....................... 96	Pigeon 102	Swallow 154
Cameroptera 170	Hadada................. 46	Pipit.....................210	Swamphen 82
Canary 212	Hamerkop 48	Plover 88	Swift 118
Chat.................... 182	Harrier 64	Pochard................ 38	Tchagra 142
Cisticola 164	Harrier-Hawk 58	Pratincole 96	Teal.......................36
Cliff Chat 184	Hawk....................56	Prinia 168	Tern 98
Coot 82	Hawk-Eagle............70	Prion 40	Thick-knee 84
Cormorant..............54	Helmetshrike 138	Puffback 140	Thrush................. 178
Coucal 106	Heron 50	Pytilia 200	Tinkerbird............ 130
Courser.................. 96	Hobby 74	Quail.....................32	Tit........................ 148
Crake 78	Honeybird 132	Quailfinch 204	Tit-Flycatcher 186
Crane 82	Honeyguide............ 132	Quelea 196	Trogon 120
Crombec............... 160	Hoopoe 126	Rail 80	Turaco 106
Crow................... 148	Hornbill................128	Raven 148	Turnstone.............. 94
Cuckoo 108	Ibis 46	Redshank 92	Twinspot 200
Cuckoo-finch 206	Indigobird............. 204	Reeve 96	Vulture 60
Cuckoo-Hawk 56	Jacana 90	Robin...................180	Wagtail 206
Cuckooshrike 142	Kestrel 72	Rock Jumper.......... 144	Warbler............... 160
Dabchick 40	Kingfisher............. 122	Rock Thrush...........184	Wattle-eye 136
Darter....................54	Kite 58	Roller 120	Waxbill 202
Dove 102	Knot 94	Ruff 96	Weaver 192
Drongo 146	Korhaan 76	Sanderling............. 94	Wheatear............. 182
Duck 34	Lapwing 86	Sandpiper.............. 92	Whimbrel 90
Eagle 70	Lark 150	Saw-wing............. 154	White-eye 174
Eagle, African Fish...56	Longclaw 208	Scimitarbill 126	Whydah 206
Eagle-Owl 114	Malkoha 108	Scrub Robin 180	Widowbird 198
Egret..................... 50	Mallard 36	Secretarybird 58	Woodpecker.......... 134
Eremomela 170	Mannikin 204	Seedeater 212	Wren-Warbler........ 170
Falcon....................74	Martin 154	Shearwater 40	Wryneck 132